普通高等教育创新型人才培养系列教材

U0204304

Modern Aircraft Electrical System
现代飞机电气系统

Zhou Jiemin   Tao Siyu

周洁敏   陶思钰   编著

北京航空航天大学出版社

Abstract

This book mainly introduces not only the composition, structure and operation principles of future advanced aircraft, More-Electric Aircraft/All-Electric Aircraft (MEA/AEA) but also the technical characteristics of MEA. It is specially written for scientific research staff and engineering personnel engaged in aircraft technology, and it systematically introduces the development overview of MEA/AEA, architecture of MEA/AEA, power supply system of MEA, power distribution system of MEA, electric actuating technology of MEA, environmental control system of MEA, electrical lighting technology of MEA, brake system of MEA, thermal management and energy optimization of MEA, common core system of MEA and application of power electronics technology of MEA, etc.

This book can be used as a teaching material of subject development course for students engaged in aircraft business, as well as a reference book for scientific research and technical personnel and engineering technicians.

图书在版编目(CIP)数据

现代飞机电气系统 = Modern Aircraft Electrical System：英文 / 周洁敏,陶思钰编著. -- 北京：北京航空航天大学出版社,2024.5

ISBN 978-7-5124-4412-6

Ⅰ.①现… Ⅱ.①周… ②陶… Ⅲ.①飞机—航空电气设备—英文 Ⅳ.①V242

中国国家版本馆 CIP 数据核字(2024)第 099434 号

Modern Aircraft Electrical System
现代飞机电气系统
Zhou Jiemin   Tao Siyu
周洁敏   陶思钰 编著
策划编辑 龚 雪   责任编辑 孙兴芳

*

北京航空航天大学出版社出版发行
北京市海淀区学院路 37 号(邮编 100191)   http://www.buaapress.com.cn
发行部电话:(010)82317024   传真:(010)82328026
读者信箱: goodtextbook@126.com   邮购电话:(010)82316936
北京富资园科技发展有限公司印装   各地书店经销

*

开本:787×1 092   1/16   印张:17.5   字数:448 千字
2024 年 8 月第 1 版   2024 年 8 月第 1 次印刷   印数:1 000 册
ISBN 978-7-5124-4412-6   定价:69.00 元

Preface

By the 2020s, the aircraft transport market was estimated to grow at 5% per year, a competition growing mainly between the US and Europe, and new commercial demand emerging in North America and Asia. With the requirements of "carbon emissions", "carbon peak" and "carbon neutrality" in energy applications, the aviation industry will face two major economic and environmental problems, and it is estimated that 2% of carbon dioxide emissions into the atmosphere are from the civil aviation industry. It is roughly estimated that saving 1 kg of fuel per flight is equivalent to saving 1 700 tons of fuel and reducing 5 400 tons of carbon dioxide emissions per year for the entire air transport industry. Further more, large savings will lead to large profits. So the aircraft industry is focusing on new technologies for the entire cost (design, development and manufacturing) and fuel consumption.

The concept of More-Electric Aircraft/All-Electric Aircraft (MEA/AEA) has been proposed for a long time. With the technological development of high-power electronics and distributed structures, MEA/AEA gradually began to be used in aircraft. The MEA application began with military aircraft and expanded back to civilian aircraft. The technology was first used in flight control braking and reverse thrust systems, and then successfully applied in electric environmental control, electric fuel pumps, and magnetic levitation bearings for engines and generators.

MEA/AEA can make good use of electricity on the aircraft and reduce fuel consumption. Its emergence has not only changed many systems on the aircraft into electric power systems, but also integrated and transformed the power generation and distribution systems on the aircraft from the whole machine.

With MEA/AEA, some complex, heavy, maintenance and easy damage can be eliminated, and working at high temperature and high pressure and flammable hydraulic oil along with some hydraulic pipelines, pumps and valves can also be removed, so that the weight of the aircraft is reduced a lot. MEA technology makes military aircraft more mobile, reducing the possibility of being shot down by enemy gunfire, and reducing operating costs for civilian aircraft. In principle, MEA can use the energy on the aircraft more effectively and reasonably. The engine no longer provides power for hydraulic and pneumatic power, and the power can be extracted from the low speed rotor, not from the high speed rotor together with hydraulic and pneumatic power, so the engine operation is more effective and reasonable.

The electric Environment Control System (ECS) can greatly improve the performance of the engine, because it does not directly interfere with the work of the engine, and the engine compressor can be designed independently; another advantage is that it can make the engine bypass reletively high, because the airflow required by ECS is no longer drawn from the engine high pressure compressor, but another compressed air source, greatly reducing the fuel consumption. The ECS also keeps the minimum engine thrust unlimited by environmental control requirements, so the aircraft can fly on a more fuel-efficient flight profile. In addition, MEA/AEA technology has the advantages of low cost, high reliability, flexible design and good performance.

The MEA is also a challenge for the aviation industry, where aircraft manufacturers need to integrate electric motors and control devices with their own systems and equipment. Due to the dramatic increase in electricity consumption, the capacity requirement of generators is higher. For example, the B787 has used 1.45 megawatts, and each generator has a capacity of 250 kV·A. The power demand of environmental control system has reached 500 kV·A; the capacity demand of brake system has reached 400 kV·A, and the capacity demand of ice has reached 100 kV·A.

MEA technology is moving forward to adopt advanced small components, reduce costs and improve efficiency. And enabled the aircraft to use more MEA/AEA technology, eventually developing the aircraft into an all-electric aircraft.

In view of the above situation, it is very urgent to comprehensively carry out the scientific research and teaching of MEA/AEA. For more than 20 years, the author has carried out the teaching and scientific research work of aircraft electrical system, and published the corresponding textbook—*Aircraft Electrical System* (Science Press, 1st. published in 2010, 2nd. republished in 2018, 3rd republished in 2023).

Since 2013, the course of "Discipline Development Platform" was launched in higher teaching requirements. In order to meet the needs of teaching reform. The course of "More-Electric Aircraft Electrical Technology" was opened, the course construction was started, and data collection, teaching research, courseware and website production, video shooting have been carried out.

Since 2012, with the development and demand of domestic international students' teaching, the course of "Aircraft Electricity System" has been opened and a large number of foreign language materials have been collected. In view of the above situation, it is necessary to compile this textbook.

This book focuses on the analysis, summary and induction of the relevant technologies of MEA/AEA, which involve a wide range of fields and knowledge. Different groups of readers are completely different, and have a wider understanding of more cutting-edge knowledge, which is consistent with the current advocated cultivation purpose. In addition to the teaching requirements, the needs of self-study are also taken into account in the aspect of compilation system and narrative methods.

The selected contents in this book are applicable to the reference of scientific research, technical personnel, related scientific and technological personnel engaged in aircraft research and production departments, and this book can also be used as a professional teaching reference book for undergraduate and graduate students of aviation electrical engineering, civil aviation electronics and electrical engineering, electrical electronics and electrical automation. Some of the electrical and electronic terms and units of measurement involved in this book strive to comply with the documents issued by the International Commission on Metrology and the State Administration of Technical Supervision.

The applicable teaching class hours of this book are 32—40 class hours. Due to the limitations of the author's experience and level, there are inevitably shortcomings or mistakes in this book. I expect readers to criticize and correct them.

Zhou Jiemin

In March, 2024

Nanjing University of Aeronautics and Astronautics

Contents

Chapter 1 Introduction

The impressive achievements in the field of aviation with the advancement of energy-efficient aviation technology have led to a dramatic change in the architecture of modern aircraft, with fundamental changes in the handling and design methods of engines, generators and various electrical equipment.

The European Airbus A380, the B787 and US Lockheed Martin F35 are representatives of modern advanced aircraft. These three new aircraft are aircraft that apply the latest modern technology, also known as More-Electric Aircraft/All-Electric Aircraft(MEA/AEA).

1.1 Definition of MEA/AEA and Changes in Power Consumption

Because electrical energy is clean, quiet, easy to transmit and automate, almost every system in the aircraft that accomplishes the mission attempts to use electrical energy as an operating energy source, and efforts are made to replace other energy sources with electrical energy.

1.1.1 Definition of MEA/AEA

The aircraft that uses electrical energy to replace the original hydraulic, pneumatic and mechanical energy, and uses as much electrical energy as possible for most of the secondary energy sources is called an MEA, which is the representative of modern advanced aircraft. The aircraft that uses electrical energy to operate and drive all of its electrical systems, including combat effectiveness control system, and uses electrical energy to distribute all of its secondary energy, is called an AEA.

1.1.2 Changes in Electric Power Consumption

Fig. 1.1.1 shows the development of aircraft on-board power, with the MEA representing the A380 and the nearly AEA representing the B787 showing a significant increase in power used compared to the commercial aircraft of the past.

As can be seen from Fig. 1.1.1, the A380 AC main power capacity has reached 600 kV · A because it is equipped with 4 engines, each of which has a 150 kV · A AC main generator. And the B787 main power capacity reaches 1 000 kV · A because it is equipped with 2 engines, each of which has 2 generators, and the installed capacity of the generator up to 250 kV · A.

As Fig. 1.1.2 shows the power consumption versus the weight of the aircraft, it is clear that the B787, an MEA with a significantly higher installed capacity, equivalent to around 10 times that of a conventional aircraft, has the greatest power to weight ratio, i.e. the most energy efficient.

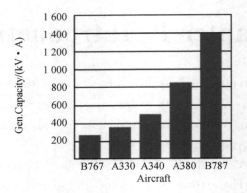

Fig. 1. 1. 1   Growth in electrical power used by aircraft

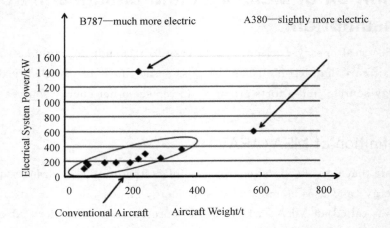

Fig. 1. 1. 2   Power consumption versus aircraft weight

1.1.3   Advantages of MEA/AEA

Most modern aircraft use an Integrated Drive Generator (IDG) for power generation. The generator drive is used to convert the varying engine input speed to a constant speed to the generator spindle via a Constant Speed Drive (CSD). The power output of the IDG is mainly limited by the maximum power of the CSD. The concept of an AEA is that all loads of the aircraft operation with electrical energy.

The main function of the Power Supply System (PSS) is to convert the mechanical power from the engine shaft into various types of energy, such as electrical, mechanical, optical, electromagnetic, thermal, and chemical energy, etc., and to distribute it throughout the aircraft. The characteristics of an MEA/AEA are discussed from several aspects below.

1. Overall Aircraft Performance Impact

The implementation of the AEA concept will eliminate hydraulic and pneumatic systems and associated equipment to improve the performance of medium-range aircraft, notably by:

① Reduction in fuel consumption by 8%—10%.

② 6%—7% reduction in takeoff weight.

③ Reduction of direct maintenance costs by 4%—6%.

④ Reduction of life cycle costs by 4%—5%.

2. Electrical Energy Conversion in Power Supply Systems

The increase in the power of aircraft electrical equipment requires the development of power supply systems with a generator capacity from 150 kV · A to 250 kV · A. The development of high power IDG with high reliability and low maintenance costs has led to the discontinuation of the use of IDG due to the enormous difficulties faced.

B787 with 250 kV · A (supplied by Hamilton Sunderstrand) and A380 with 150 kV · A (supplied by Thales) varying frequency generators operate in the speed range of 10 800 to 16 000 r/min. The adoption of the MEA/AEA design concept on aircraft such as the A380 and B787 has improved operational and economic performance, in particular by reducing fuel consumption per passenger by 17%—20%. The PSS includes the following electrical energy converters:

① A power generation system converts mechanical energy from the engine shaft and chemical energy from the power source into electrical energy.

② The distribution system converts the electrical energy generated by the power generation system to meet the current, voltage and transient characteristics of the grid.

③ The actuator system converts electrical energy into another form of energy, such as mechanical, optical, electromagnetic, thermal, etc.

3. Major Advantages of MEA/AEA

① Low heat generation by electric actuation, with good fault tolerance.

② Reduction of energy types.

③ Electric drive systems are superior to hydraulic drive systems.

④ Weight reduction.

⑤ Easy installation of aircraft systems.

⑥ The MEA facilitates the implementation of self-tests, easy maintenance and simplified ground support equipment.

4. Disadvantages of Power Generation Systems

The power generation system of the next generation of advanced aircraft is a centralized power generation system, and the functional components of the power generation system are managed by an integrated control system.

Centralized power generation system needs more power conversion devices, because most loads cannot be directly supplied by the main power supply, so power conversion is needed. Another method is to use a decentralized PSS that does not adjust the main generator. Various controls and regulation are power converted simultaneously by the distribution system and the driving element to obtain the required power supply voltage level, type and quality.

The power supply is a variable frequency power supply and therefore its implementation, control and monitoring of the distributed distribution device in the PSS operate under variable frequency conditions.

More electric generators are open-loop systems without generator regulators, which means that there are no abnormal patterns and transient deviations caused by regulator operation, and no need for real-time control of system parameters, but only the electrical loads. Despite the many benefits of more-electric Permanent Magnet Generators (PMGs), the main disadvantages of this decentralized distribution system are the particularly high demands placed on the electrical loads and the distribution system, and the inability to demagnetize the generator in an emergency.

1.2　Key Technologies Needed for MEA/AEA

Based on system-level optimization technology, the systems will use newer technology to meet the overall performance of the aircraft, bringing about a revolution in MEA/AEA electrical technology, with the more prominent key technologies being:

① Changes in aircraft architecture.

② Aircraft integral start-up/power generation technology.

③ Varieties of aircraft power systems, power handling methods.

④ Aircraft redundant power distribution technology.

⑤ Aircraft Environmental Control Systems(ECSs) and electrical anti-icing technology.

⑥ Aircraft power telegraphy and actuation technology.

⑦ Integral aircraft takeoff and landing systems.

⑧ Aircraft public management system.

Breakthroughs in these key technologies provide technical support and security for the transition from MEA/AEA.

1.3　Current Status of Advanced Aircraft Electrical Technology

In the highly competitive world of aviation, reducing the maintenance costs of aircraft is an important measure to reduce the life-cycle costs of an aircraft. Aircraft manufacturers are placing greater emphasis on new technologies to reduce aircraft flight and maintenance costs, from optimizing the performance of the aircraft's subsystems to optimizing the aircraft's performance in the context of the aircraft as a whole. MEA/AEA technology is a concept that has been developed based on the need to optimize the design of the entire aircraft power train, revolutionizing the aircraft power train. The MEA/AEA uses electrical systems instead of other secondary energy sources on the aircraft as far as possible. This includes hydraulic systems, pneumatic and wind energy systems, which can significantly reduce system weight and cost, and improve aircraft maintainability and reliability.

In the late 1990s, the US Air Force's MEA programme and Europe's Totally Integrated More-Electric System (TIMES) programme for civil aircraft both explored an optimizing aircraft design by considering the aircraft as a whole. The Power Optimization Aircraft (POA) programme aims to recognize, optimize and validate innovative aircraft equipment systems and aircraft system architectures to reduce peak non-propulsive power consumption by 25% and fuel consumption by 5%, while maintaining the cost and reliability of the equipment.

Fig. 1. 3. 1 illustrates the future growth of electrical power of MEA. As aviation technology evolves, there will be progressive applications of advanced technologies, such as no compressed air extraction systems, no hydraulic systems, the birth of More-Electric Engine(MEE) and the use of fuel cell, eventually leading to fully electric aircraft. Thus when power electronics, advanced materials and thermal management technologies become possible, there is the possibility of achieving an AEA.

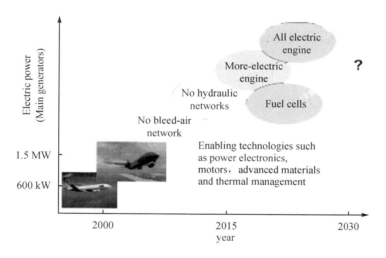

Fig. 1. 3. 1   Future growth in electrical power of MEA (published in 2000)

1.3.1　Development Status of Aircraft Electrical Technology

Aircraft electrical technology is facing profound changes in order to meet the development of MEA/AEA, placing new demands on the development of aircraft electrical technology.

1. Power Capacity Requirements

It is expected that the power demand for aircraft airborne power supply will increase greatly in the future, and each engine will provide 500 kV • A of power. It is mainly used for civil aircraft flight and entertainment equipment, information service equipment, passenger comfort equipment and High-Energy Laser Weapons(HELW) on military aircraft, as well as aircraft Flight Control Electric Actuator (FCEA), electric brake, Electric Environment Control System(EECS), electric deicing and other new electrical equipment. It is estimated that the electrical environmental control system of the 350-seat passenger plane requires about 400 kV · A, and deicing also requires 150 kV · A. In the European POA program for a single capacity of more than 300 kV · A, the US military is also studying 500 kV · A Starter/Generator(S/G) to meet the demand for power capacity.

2. Changes in the Power Supply System

As the capacity of power systems increases, the application of 28 V DC Power Supply System (PSS) on aircraft will be further reduced and at this stage is mainly based on 115 V/400 Hz AC PSS. Due to the efficiency and reliability of the constant frequency system, it will rarely be used in large capacity systems in the future. Depending on the engine speed variation range variable frequency, AC frequency range is generally 360—800 Hz, and the A380 and B787 have used this Variable Frequency AC Power Supply System (VF AC PSS). In order to reduce the weight of the transmission cable, higher voltages such as 230 V AC or 115 V/200 V are required to obtain 270 V DC through three-phase full wave rectification.

3. Uninterrupted Power Supply Requirements

For aircraft that rely on electrical actuators, the requirement is to keep the flight control actuators powered throughout the flight envelope. Although short interruptions are permitted, however, MEA/AEA with fully electric flight control will never allow long power interruptions, so the Fan Shaft Driven Emergency Power Supply System(FSDEPSS) was introduced in the European Totally Integrated More-Electric System (TIMES) programme to provide continuous power to the aircraft during all phases of flight and in Emergency Condition(EC), with the F35's Integrated Power Unit (IPU) being one of the main power sources in both normal and EC.

4. Automatic Load Management

With the increase in electrical loads on aircraft, automatic management of loads can only be achieved by computer. Control of loads, condition detection, fault isolation and system reconfiguration via the Solid-State Power Controller(SSPC) and the Mechanical Electronic Power Controller(MEPC) are essential requirements for Multi-Power/All-Power Aircraft (MPA/APA) electrical technology.

1.3.2　The Foundation to Support Technology Requirements

The key technologies of MEA/AEA break the traditional concept of aircraft systems, strengthen the integration of multiple disciplines and reflect the innovative research results in various fields at present, thus putting forward higher requirements for the research of basic materials, components and system control technologies. The key technologies and future advanced development trends of MEA/AEA are shown in Fig. 1. 3. 2. In the development of MEA/AEA technology, the following basic supporting technologies need to be particularly strengthened.

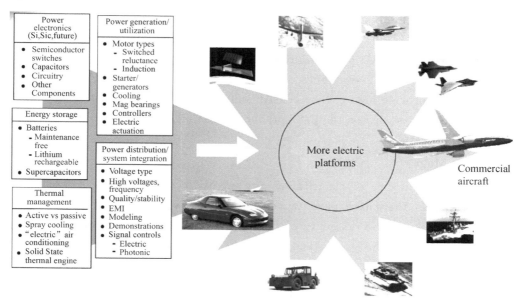

Fig. 1. 3. 2   Key technologies and future advanced development trends of MEA/AEA

1. High Temperature Power Electronics

The Si-based devices are already quite mature, and the structural design and process are already quite well developed and close to the limits of material properties. The breakdown field strength of SiC devices is 10 times that of silicon devices, and the thermal conductivity is 3 times that of SiC devices, and the high breakdown field strength can make the doping area of SiC power devices thinner and the doping concentration greater, which reduces the through-state resistance and can greatly reduce the through-state and switching losses, and can increase the operating frequency of the devices; good thermal conductivity can make the capacity of the devices greater, and the maximum operating temperature may exceed 600 ℃. In aerospace, high-temperature radiation environment, oil drilling, automotive electronics and radar communications etc. But it still needs to be studied in terms of reliability and practicality.

2. High Temperature Resistant Magnetic and Insulating Materials

The properties of magnetic and insulating materials change at high and low temperatures. For magnetic materials, the general operating temperature does not exceed 105 ℃, and the highest will not exceed 150 ℃. NASA Lewis Research Centre successfully researched an inverter with an operating temperature of 200 ℃; the United States Air Force Research Institute also developed 1. 5 kW, 270 V/28 V transformer in the forward converter, which could be in operation for 600 hours in 290 ℃ environment, and its magnetic material is manganese zinc material that has a Curie temperature of 310 ℃. The high temperature resistance and thermal stability requirements of the generator winding insulation material in MEA/AEA are 600 ℃. Therefore, the MEA/AEA has put forward higher requirements for insulation materials and magnetic materials.

3. Energy Storage Elements

As the capacity of aircraft power sources increases, so do the demands on the capacity of energy storage components. New types of rechargeable lithium batteries and fuel cells are being developed to provide high energy density and power at reduced weight. For example, the lithium-ion batteries used on the B787 produces five times the energy of the same weight of nickel cadmium batteries.

4. Electric Motors

With the predominance of electrical energy on MEA/AEA and the extensive use of electric mechanisms, there is a need to develop high power density and high reliability motors that will increase the reliability of the aircraft system, and the high efficiency will also reduce the weight of the system and reduce the burden on the thermal management of the system.

Power electronics and transmission: MEA/AEA require a large number of power electronics to drive the motors or to convert the power supply, and the use of conventional technologies leads to an increase in system weight and complexity in system energy management. That is why research into highly reliable and efficient power electronics and the standardization and specification of components is a key technology for the development of MEA/AEA.

5. Integrated Aircraft Systems Management Technology

The MEA/AEA will use integrated management techniques to automate the management of more than just the electrical system. The electrical system, as part of the aircraft's electro mechanical system, should also be controlled as a subsystem of the electro mechanical control system, so that the design of the aircraft can be optimized in an integrated manner from the aircraft system level. This fusion of information from multiple systems poses a huge challenge for the design of the system, which requires not only a large amount of information to be exchanged between systems, but also a greater degree of intelligence in the subsystems. The electrical system should be fully automated, with automatic load management and monitoring in normal conditions, and automatic reconfiguration of the power distribution system structure and automatic fault isolation according to the system status in fault conditions.

6. System Thermal Management Techniques

Despite efforts to improve the efficiency of the various systems and equipment, the significant increase in total power creates huge difficulties for the thermal management of the system. Thermal management technology has become a key technology for the system, especially for aircraft with a large number of composite materials, which cannot be dissipated through the fuselage and have limited space for heat dissipation, and it is more important to solve the heat dissipation problem of the electrical and other system components from the overall consideration of the aircraft.

1.4　The Development Trends for the Aircraft

The development of aircraft is determined by the advancement of engine technology and will gradually move from conventional aircraft to MEA and even fully electric aircraft, eventually realizing electrically propelled aircraft. The development trends of future aircraft are listed in Tab. 1.4.1.

Tab. 1.4.1　Trends of future aircraft

Items ＼ Aircraft category	Conventional aircraft	MEA	AEA	Electric propulsion aircraft
Model	B737, A320	B787, A380, F35	—	—
Engines	Jet	Jet engines, minimal hydraulic and pilot gas extraction	Jet engines, no hydraulic and pilot gas extraction	Electric propulsion
Hydraulic systems	Flight control, landing gear, actuators	Decrease in power	Increase in power consumption	Increase in power consumption
Pneumatic systems	Anti-icing/deicing, environmental control	Decrease in power	Increase in power consumption	Increase in power consumption
Electrical systems	Electric motors, lighting and heating, avionics, system control	Increase in power consumption	Increase in power consumption	Increase in power consumption

The power source in an electric propulsion aircraft operation with electrical energy consists of fuel cells, batteries and ground power (for ground support), which supply power to the electrical propulsion system and the aircraft systems, as is shown in Fig. 1.4.1 for the electrical energy utilization in fuel cell type aircraft.

As can be seen in Fig. 1.4.1, aircraft electrical energy is mainly used to propel the aircraft, including drive motors, lighting, heating, powering avionics, flight control actuators, landing gear, loop control systems, de-icing and controllers for various subsystems.

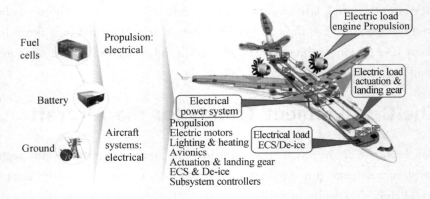

Fig. 1. 4. 1 Electrical energy utilization in fuel cell type aircraft

Aircraft will develop towards MEA, energy-optimized clean aircraft and zero-emission aircraft, while the development of engine is moving from universal More-Electric Engine (MEE) to universal open engine and universal high-speed superconducting motor. The propulsion system is developing from a built-in propulsion system to a built-in electrical propulsion system due to the development of new electrical technologies; and the engine and generator are developing from separate engines with Variable Frequency Starter/Generator (VFSG) to an integrated system with variable frequency engine starter motors.

In addition, solid state modular power modules, silicon carbide conversion, power distribution and control components, flight control using Electric Mechanical Actuator (EMA), Electrical Environment Control Systems (EECSs), Electrical Redundant Power Distribution (ERPD), dynamic energy management, high density fuel cell, high density storage, electrical landing and taxiing and solar energy utilization are all future trends. The future trends for aircraft and engines are listed in Tab. 1. 4. 2.

Tab. 1. 4. 2 Future trends for aircraft and engines

Aircraft type	MEA	Energy-optimized clean aircraft	Zero emission aircraft
Engine shape	Universal more-electric engine	General open engines	Universal high-speed superconducting motors

(**Continued**)

Aircraft type	MEA	Energy-optimized clean aircraft	Zero emission aircraft
Electricity technology	Built-in propulsion system, Variable Frequency Starter/ Generator (VFSG), solid-state modular power modules, MOS power distribution units, electrical circulation control system	Built-in propulsion system, collection of variable frequency engine starter motors, solid-state modular power modules, silicon carbide conversion, distribution and control components, flight control system with EHA/EMA, electrical circulation control system, active energy management, electrical landing and taxiing	Built-in electrical propulsion system, solid-state modular power modules, silicon carbide conversion, distribution and control components, flight control using EMA, electrical circulation control system, electrical redundancy distribution, dynamic energy management, high density fuel cell, high density storage, electrical landing and taxiing, solar energy use

1.5　Summary

This chapter introduces the definition of MEA/AEA, the changes and advantages of its power consumption, and the resulting critical technologies. It also introduces the current status of aircraft electrical technologies abroad and the considerations for the technological development in China, as well as the present state of the development of aircraft electrical technologies and the demand for the fundamental support technologies of MEA/AEA.

1.6　Exercises

1. What is an MEA?
2. What is an MEA?
3. What are the key technologies for MEA/AEA?
4. What are the basic support technologies needed to achieve an MEA/AEA?
5. What are the development trends of aircraft and engines?

Chapter 2   Architecture of MEA/AEA

With the demand for energy-saving and green aviation, the aviation industry will face two major economic and environmental problems. It is estimated that 2‰ of carbon dioxide emissions into the atmosphere come from the civil aviation industry. It is roughly estimated that about 1 kg of fuel per flight is equivalent to saving 1 700 tons of fuel and reducing 5 400 tons of carbon dioxide emissions per year in the air transport industry. Furthermore, a lot of savings will bring huge profits. As a result, the aircraft manufacturing industry is focusing on the adoption of new technologies in terms of overall costs (design, development and manufacturing) and fuel consumption.

In addition, aviation power structure, engine optimization and electrical equipment are also playing an important role, especially in reducing minimum operating costs and maximum airworthiness. Therefore, the use of aircraft energy is very important.

2.1   Aircraft Primary Energy and Secondary Energy

2.1.1   Aircraft Primary Energy

Fig. 2. 1. 1 shows the layout and utilization of aircraft energy. The equipment that provides primary energy for aircraft mainly includes engine, Auxiliary Power Unit (APU), Ram Air Turbine (RAT) engine driven generator and ground power.

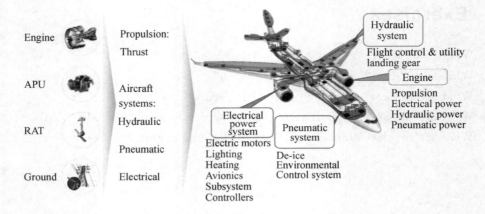

Fig. 2. 1. 1   The layout and utilization of aircraft energy

These primary energy sources provide Propulsion Thrust Power(PTP) for the aircraft and make the aircraft generate forward power. There are also hydraulic power, pneumatic power and electrical power required by various aircraft systems, which are called secondary energy.

The hydraulic power in the secondary energy of the aircraft is generated by the Hydraulic Pump (HP) directly driven by the engine, which is used for the control of the rudder surface of the aircraft, the control of the engine, the opening and closing of the hatch and the retraction of the landing gear. The pneumatic power is mainly extracted from the supercharged air of the engine, which is used for deicing and Environmental Control System (ECS) and so on. The electrical power is mainly generated by the generator driven by the engine, directly or indirectly, which supplies power to the electric motors, lighting device, heating device, avionics equipment and so on.

As early as the 1970s, the problem of unifying aircraft secondary energy into electricity was put forward. Through more than 40 years of efforts, great progress has been made in the development of MEA and AEA technology. That is the result of the failure of the hydraulic system of many aircraft at home and abroad, resulting in the destruction and death of the aircraft. The main reason is that the centralized hydraulic system has high power loss, high heating and high temperature, and the parts are easy to be damaged. The leakage of a certain part of the hydraulic system will paralyze(瘫痪) the whole system and cause irreversible faults. Based on the development of electrical materials and electrical technology, aviation scientists at that time put forward the suggestion of using EMA mechanism as the backup of hydraulic actuating mechanism, and achieved success, thus creating conditions for electrical energy to replace hydraulic energy. With the development of energy-saving engine, it is required not to extract compressed air from engine, and to use electrical energy instead of pneumatic energy. Studies have shown that extracting the same horsepower of compressed air and engine shaft power, the latter can save more fuel.

2.1.2 Energy Demand of Traditional Aircraft

As shown in Fig. 2.1.2, it is a complex network of four energy sources used by a typical aircraft. It can be seen from Fig. 2.1.2 that the architecture of a traditional aircraft has four forms of energy, namely:

① The hydraulic center provided by the engine needs to provide hydraulic power to the main flight control, secondary flight control and landing system.

② Electric power is provided by generators directly driven by the engine, and all electrical loads on the aircraft are supplied by the power distribution system.

③ The engine and APU provide pneumatic power to the environmental control system and anti-icing system.

④ The mechanical power provided by the accessory gearbox of the engine drives the generator and landing system.

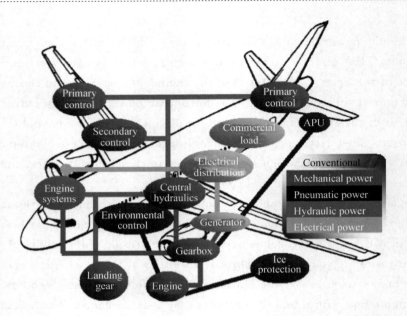

Fig. 2.1.2　A complex network shared by the four energy sources of a typical aircraft

2.2　MEA Engine

2.2.1　Summary

For a long time, it has been exploring the more-electric method for aircraft to extract power from the engine. There are usually three main ways to extract energy from the engine:

① To obtain power through the generator driven by the accessory drive case.

② To obtain hydraulic power through the engine drive pump which is also driven by the accessory drive case, but hydraulic power can also be obtained by means of electric drive or air drive.

③ The air pressure power is obtained by extracting the air from the medium-pressure or high-pressure compressor, which is used to provide energy for the environmental control system, cabin pressurisation and wing anti-icing system.

High-pressure air also provides the means to start the engine, which is taken from ground air source vehicles, APU or other engines that are already in operation.

Most civil aircraft engines are dual engines with low-pressure and high-pressure rotors. In the advanced gas turbine engine, a large part of the air bypasses engine main culvert(主涵道), and the ratio of bypass air to engine main culvert air is called the culvert ratio. Since an engine is actually a highly optimized gas generator, there is a cost in extracting the gas draw, which is not commensurate with the power being extracted. This cost becomes more severe when the engine's lane ratio increases. For example, the initial turbofan engine had a low inclusion to lane ratio of approximately 1. 4 : 1 (side culvert: engine main culvert), which in turn progressed to the engine had an approximate 4 : 1 inclusion to lane ratio. The culvert ratio of the next generation turbofan engines such as General Electric's Genex and Rollo's "Ruida" 1 000 is close to 10 : 1. Future engines tend to have a pressure ratio of (30—35) : 1, and they will be more sensitive to less and less regulation and much higher extraction of air from the central culvert of the engine.

In order to fully realize the benefits derived from emerging engine technology, a different and more efficient method of extracting power or energy for aircraft systems is necessary. The efficient extraction of energy for the aircraft, without adversely affecting the engine main lane and engine overall performance, has become a compelling reason to change the structures and techniques employed.

2.2.2 Utilization of Conventional Engine Energy

Fig. 2. 2. 1 shows the energy use of conventional engines. Conventional civilian aircraft use four energy sources: Pneumatic power can be used for environmental control and wing de-icing systems and engine start-up; electrical power can be used for avionics and electrical equipment; hydraulic power can be used for pump drive and flight control action; mechanical power from the engine accessory gearbox is used to drive the hydraulic and fuel pump and so on.

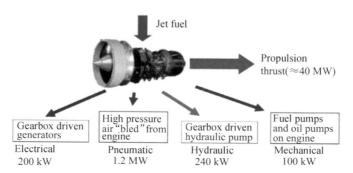

Fig. 2. 2. 1 Utilization of conventional engine energy

As can be seen in Fig. 2. 2. 1, a generator driven by a gearbox drive is required to deliver 200 kW of electrical power; the pneumatic power drawing from the engine's high pressure is 1. 2 MW power; the use of a gearbox to drive the hydraulic pump at 240 kW power; the fuel pump and slide pump on the engine operate at a mechanical power of 100 kW. The rational use of energy has led to the design of MEA that must replace conventional systems with electrical ones.

2.2.3 Energy Utilization of MEE

The energy utilization of modern advanced MEE is shown in Fig. 2. 2. 2.

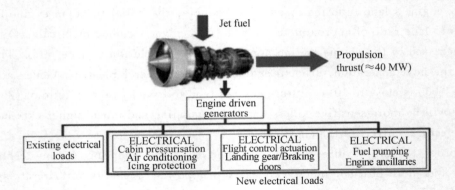

Fig. 2. 2. 2   Energy utilization of modern advanced MEE

As can be seen from Fig. 2. 2. 2, the MEE uses only generators driven by the engine to generate electrical energy. Mainly power supply has been installed electrical load, including three kinds of electrical load. First, it is used for electrical load such as cabin pressurisation, air conditioning, icing protection, etc. Second, it is used for electrical load such as flight control actuation, landing gear/braking doors. Third, it is used for electrical load such as fuel pumping or engine ancillaries.

Fig. 2. 2. 3 is a comparison diagram of the power extraction between conventional aircraft and MEA. On the left is the conventional power extraction scheme for applying gas diversion, and on the right is the modern advanced MEA engine scheme, which is called MEE. In conventional engines, it is necessary to extract air from the engine for engine anti-icing, wing anti-icing, environmental control and turbo charging and so on. Through the drive shaft installed in the lower part of the fan shell to drive the accessory drive box, the

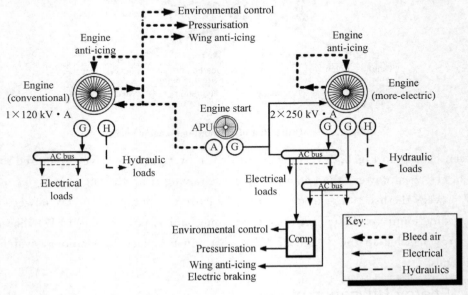

Fig. 2. 2. 3   Comparison diagram of the power extraction between conventional aircraft and MEA

power extraction of the engine is realized, and the power is provided to the central system by the generator and the hydraulic pump. Compared with the conventional aircraft engine, the power extraction method of the MEE is completely different. It is no longer extracted from the engine high pressure stage and no longer affects the thrust of the engine. Only the gas extraction is extracted for ice prevention in the engine inlet.

In an MEE, the air intake of the engine fan is used for the engine anti-icing, and there is no air intake into the anti-icing system. The MEE components of the "Ruida" 500 are shown in Fig. 2. 2. 4.

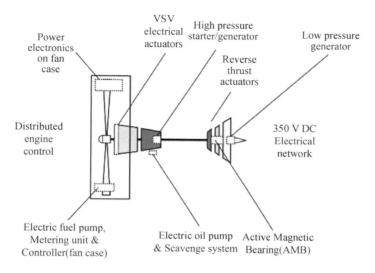

Fig. 2. 2. 4   MEE components of the "Ruida" 500

The European Power Optimization Aircraft(POA) program joint group has verified the MEE "Ruida" 500. It has the following characteristics:

① The high voltage starter/generator is a Permanent Magnet Generator (PMG) providing a power of 150 kV · A.

② The low-voltage Fan Shaft Drive Generator(FSDG) is a Switch Reluctance Generator (SRG) that provides 150 kV · A power supply.

③ The Power Electronics Module (PEM) is located on the engine fan housing to provide 350 V DC power to the engine and multiple electric components of the aircraft.

④ The Electric Fuel Pump Measurement System(EFPMS) includes motor, pump and electronic device. The total electrical power used is about 75 kW, which can extremely accurately measure the flow. The pump only provides the required fuel flow, so it does not waste pressurized power and no subsequent heat dissipation problems.

⑤ Electric lubricating oil pump.

⑥ Electrical actuators that can be used for various purposes, such as adjusting the action of the flow blade, instead of fuel pressure action using EMA. Two physically identical actuators are used to form an active/driven structure form, namely the reverse thrust actuator, or the spiral linear actuator.

⑦ Active Magnetic Bearing (AMB).

⑧ The power supply voltage of the engine is 350 V DC.

As shown in Fig. 2. 2. 5, it is the main electrical components of the more-electric "Ruida" 500. The generator of that High Pressure Starter/Generator (HPSG) is a permanent magnet generator, which can generate 150 kV · A electric power; switched reluctance generator located on the fan housing can provide 150 kV · A electric power. FSDG has the important feature of providing high power in emergency situations. Under the condition of self-rotation, the fan shaft of the engine will continue to rotate at 8% of the total engine speed, so a considerable amount of electric power can still be extracted from the FSDG by using the flexibility of switched reluctance motor. Therefore, FSDG provides a device to replace RAT, which supplies emergency power and has the advantage of being integrated with the engine. Another advantage is that FSDG can always be used, while RAT is a one-off emergency system, which may not work when needed. The PEM installed on the fan frame of the engine can provide 350 V High-Voltage Direct Current (HVDC) to the engine and aircraft electrical components.

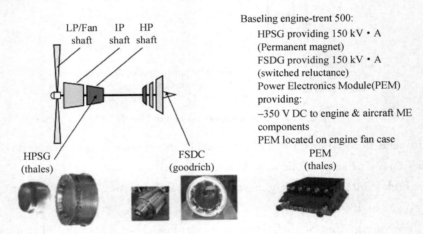

Fig. 2. 2. 5　The main electrical components of the more-electric "Ruida" 500

In addition to the increased electric actuation mentioned above, the main electrical features of the engine are shown in Fig. 2. 2. 6.

The busbar voltage of the engine is 350 V DC. The High Pressure Starter/Generator (HPSG) receives 350 V DC (Spanish National Grid) to start the engine from the outside. Once the engine is running, the HPSG provides 350 V DC to the engine busbar through the power electronic module for use by other subsystems, such as fuel metering, AMB and actuators. Once the engine is running, the FSDG will also supply power to the engine power bus 350 V DC.

The distributed engine control adopts definite Controller Area Network (CAN) bus, the

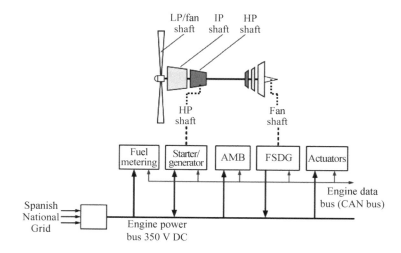

Fig. 2. 2. 6 An overview of the electrical structure of the more-electric "Ruida" 500

highest speed is up to 1 Mb/s and the lowest speed is 125 kb/s, which is used to realize the comprehensive control function. High-speed CAN bus is used to integrate the main engine control functions such as Variable Stator Vane (VSV), fuel metering, engine electronic control and lubricating oil system. Low-speed CAN bus is used to control secondary control functions such as low-voltage generator, AMB, lubricating oil return pump and lubricating oil ventilation device.

2.3 Benefits of Secondary Energy Unification

Energy is indispensable for airplanes to fly in the air. Fuel (chemical energy) is usually called primary energy, which is produced after ignition and combustion of the engine and is used to propel the aircraft forward and to work on the equipment on board. This part of energy used for the work of on-board equipment is usually in the form of electric power, pneumatic power, hydraulic power and mechanical energy, which is called secondary energy. They are transmitted to the machine equipment to work in the form of generators, hydraulic pumps, high pressure air suction equipments and motors installed on the engine accessories. Secondary energy is unified to work for electrical energy, or from aircraft to MEA and AEA, which brings greater progress in aircraft technology and promotes the development of aviation science and technology. It helps to realize aircraft safety, reliability, economy and green energy-saving aviation, and it is also the only way to realize "carbon emission and carbon neutralization". It reduces the burden of aircraft maintenance and reduces the operating cost of full-life aircraft.

2.3.1　Benefits of Engine Structure Optimization

The simplification of the structure of the engine increases the Thrust to Weight Ratio (TWR) of the engine, reduces the upwind area of the engine, and saves the fuel consumption of the engine.

In order to drive a lot of accessories, traditional engines must have a special accessory casing, which is equipped with fuel pumps, lubricating oil pumps, hydraulic pumps, generators and starters. In the early days, jet engines used electric starters or generators, and some were equipped with compressed air starters, but now most of them are gas turbine starters. The installation of these accessories greatly complicates the structure of the engine and increases the upwind area, resulting in an increase in upwind resistance. The MEA/AEA uses the engine built-in starter/generator, and the motor and the engine are on the same axis, no longer driving fuel pump, lubricating oil pump and hydraulic pump, thus canceling the accessory casing and significantly simplifying the structure of the engine. Because the compressed air of the engine is no longer extracted, the efficiency of the engine is improved. The rotor of the engine can be suspended, that is, the all-electric engine technology has also entered the stage of development and verification. The suspension of the engine rotor can increase the rotor speed, cancel the lubricating oil pump and lubricating oil system, further simplify the structure and improve the performance. As shown in Fig. 2.3.1, the structure of a traditional engine is compared with that of an MEE.

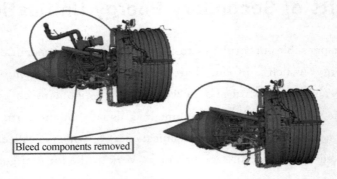

Bleed components removed

Fig. 2.3.1　Schematic diagram of structure comparison between traditional engine and MEE (B787 APU)

As can be seen from Fig. 2.3.1, the structure of the traditional engine on the left is much more complex than that of the MEE on the right, mainly because the high pressure of the latter causes the pipeline to be removed. The APU of the air vent line is removed, as shown in Fig. 2.3.2.

After removing the air intake from the APU, the components removed are: load compressor, Inlet Guide Vane (IGV) mechanism and actuator, Surge Control Valve (SCV), exhaust line and exhaust sensor, resulting in significantly improved reliability.

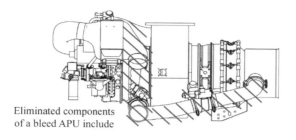

Eliminated components
of a bleed APU include

Fig. 2. 3. 2   The APU of the air vent line is removed

2.3.2   Simplifies the Internal Structure of the Aircraft

The hydraulic system, like the electrical system, is distributed in all parts of the aircraft. Due to the crisscross of pipelines and circuits, it is difficult to use and troubleshoot. When in use, the hydraulic system is a large heat source, and must be cooled using fuel, further complicating the structure. Fig. 2. 3. 3 shows a schematic diagram of pneumatic ducting and associated systems eliminated of B787.

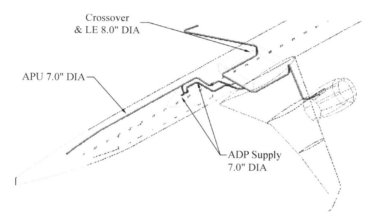

Fig. 2. 3. 3   Elimination schematic diagram of inflatable pipeline of the B787 and its related system

Due to the simplification of aircraft architecture, the following benefits have been brought:

① The energy of the engine can be extracted according to the electrical energy demand.

② More efficient power extraction and engine cycle greatly reduce the weight.

③ The number of engine and APU parts is reduced, and the exhaust system, pipeline leakage and overheat detection are eliminated.

④ The reliability is improved, which is reflected in the use of highly reliable power electronic products, which reduces the number and complexity of engine and APU components.

⑤ Through the elimination of pneumatic devices, the maintenance cost is reduced, which is mainly reflected in the elimination of pipes and valves and the cancellation of overheat detection.

As shown in Fig. 2. 3. 4, the performance of MEE with and without engine pressure energy extraction is compared. In Fig. 2. 3. 4, the energy utilization of B767 – 400 and B787 is compared under various flight profiles, such as ground taxiing phase, take off, climbing, cruising, approach and so on. Obviously, the B787 can greatly save energy compared with the conventional jet B767 – 400.

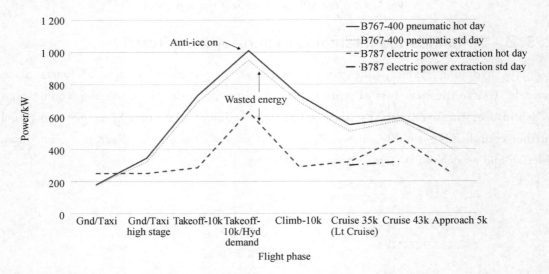

Fig. 2. 3. 4　Comparison of the performance of MEE with and without engine pressure energy extraction

2.3.3　Simplified the Ground Support Equipment

The unification of aircraft secondary energy not only helps to improve the reliability, maintainability, ground support performance and viability of aircraft, but also reduces the weight of aircraft. The following is an example of the ground support of B787.

As shown in Fig. 2. 3. 5, it is a schematic diagram of the external power connector of B787, in which a single ground power connector with a capacity of 90 kV · A meets the power supply of the hatch door service. The B787 has four 90 kV · A ground power connectors, with two external power connectors in the front and two in the middle, all of which are main external power connectors, which is an off-ground power supply for airborne power supply system.

1. Main External Power Connector

The function of two 90 kV · A ground power connectors in front area:

① Typical functions are for opening and closing doors, such as display power supply on the flight deck, refueling power supply, cargo and baggage compartment operation, and kitchen services, including cooling.

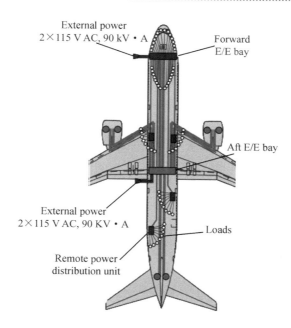

Fig. 2.3.5 Schematic diagram of B787 external power interface socket

② The operating power supply of part of the nitrogen generation system.

③ Basic cabin services, such as lighting of all cabins; partial ventilation (toilets and kitchens), maintenance of some internal equipment; power supply of some kitchen equipment (without ovens), but does not support cargo hold heating; hydraulic power supply.

2. The Function of Front 90 kV · A Ground Power Connector

The main functions of a 90 kV · A ground power connector in the front of the aircraft are: flight cabin display; refueling; control power supply for cargo and baggage compartments; kitchen power supply, including cooling; basic cabin services, such as partial cabin lighting, partial ventilation (toilet and kitchen). It can not be used for nitrogen generator, internal equipment maintenance, kitchen electricity, cargo hold heating, hydraulic power supply.

3. Two APU Generators

Two APU generator power supplies, each with a capacity of 225 kV · A and a total of 450 kV · A, can be used for the service of ground power supply. The main applications of APU power supply are: Power supply for the Environment Control System (ECS), power supply for the cargo and luggage compartments, electric powered by the cargo compartment heater, nitrogen generator power supply; power supply of hydraulic system; power supply for kitchen facilities, including refrigeration; power supply for basic services, such as all cabin lighting, cabin ventilation, internal equipment maintenance, power supply for all kitchen facilities, refueling power supply and ground power supply, etc.

4. Other Ground Power Supply Applications

(1) Four Ground Power Sockets

In addition to the above ground power sockets, there are two 90 kV · A ground power sockets in the front and middle of the aircraft. As shown in Fig. 2. 3. 6, there are two electrical connectors at the front of the ground power service vehicle. Attention should be paid to the scope of use:

① Limited to engine start-up power supply or other unique maintenance operations.

② All functions are covered by single power supply and dual power supply.

③ Limited to some power supply and testing, that is, airborne ECS, electronic cargo tank heating, operation of nitrogen generation system and hydraulic power supply.

Fig. 2. 3. 6　Schematic diagram of the connection of the ground power service vehicle

(2) Traction Locomotive Operation Power Supply

The locomotive used for B787 traction needs power supply, and B787 has two traction locomotives, which need to be close to the electrical capacity of 10 kV · A.

(3) Power Supply of Battery Operated by Towing

Traditional towing uses APU hydraulic brake, while B787 eliminates hydraulic energy and uses electric brake. Towing battery power supply can only be used for braking during the day, and the application of storage battery to night lighting and brake power supply is under study.

From the above analysis, for the B787, the existing airport infrastructure is also applicable, and the existing ground service equipment such as power vehicles is also applicable. A single 90 kV · A power supply supports traditional operation power supply, and the increase of ground power supply capacity is used to improve the power supply capacity. When the power supply capacity is lower than the 10 kV · A, the aircraft still accepts and completes some ground operations. The possibility of towing aircraft powered by battery is being studied.

Now AEA technology has been developed to the field of automobiles, tanks and ships. Electric vehicles, all-electric tanks and electric-driven ships have become the goals of future development, and related work is under way.

2.4 MEA Hybrid Power System

In the past few decades, the aviation industry involving military and civil aircraft has made considerable efforts and contributions to the development of MEA transportation system, giving birth to many research projects based on the concept of MEA. MEA uses electrical energy to drive the aircraft subsystem and electrical energy drives to replace systems driven by hydraulic and pneumatic energy. As the electrical energy operating institutions can reduce the environmental impact and improve the overall performance, it is particularly necessary to pay attention to the management of airborne energy from a system point of view, including the characteristics of airborne power supply, storage, distribution and consumption.

Another feature of MEA is that it increases the flexibility due to the distribution of electrical energy at the load end, which can reduce the quantity and quality of cables, reduce the loss of power, and significantly reduce the weight. The PSS of MEA is easy to be reconstructed, for example, when the load is faulty or overloaded, it can be realized by turning on or disconnecting the power switches of different buses, which provides more choices for fault diagnosis. The A380 and B787 are commercial transport aircraft equipped with large-capacity generators, widely using technologies related to the concept of MEA, such as from generators to power electronic equipment, from energy storage to the control of the entire electrical system.

This section mainly briefly introduces the More-electric Propulsion System(MEPS).

2.4.1 More-Electric Propulsion System

In order to meet the newly proposed challenges of MEA and MEPS in the aerospace field, the use of more-electric energy in the aerospace field is a key change in the development of aircraft in the future. Through the continuous investment in the research and development of MEA and advanced propulsion systems, we can reduce carbon dioxide and nitrogen oxide (CO_2/NO_x) emissions, reduce noise, reduce flight costs and reduce life-cycle maintenance costs. One of the key factors for the realization of the MEPS is the technological breakthrough of more-electric generators and power electronic converters, so that the gas turbine engine can be updated.

Fig. 2.4.1 shows the development curve of SAFRAN on MEA and MEPS. It can be seen from the figure that the level of electric power consumption increases exponentially. The voltage level is less than 1 000 V to greater than 1 000 V.

Because the junction temperature of high temperature Wide Band Gap (WBG) power semiconductors is more than 200 ℃, its introduction promotes the application of high power density power electronic equipment in harsh environment temperature, and has a great impact on the aerospace field, especially in MEA and MEPS. The volume and weight of

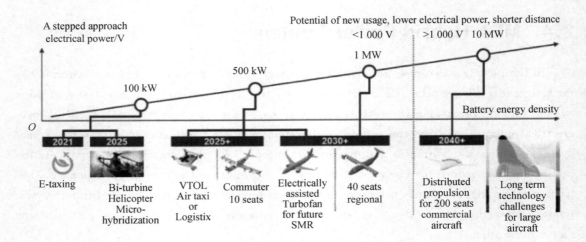

Fig. 2. 4. 1　The development curve of SAFRAN on MEA and MEPS

power electronic systems can be reduced and integrated with gas turbine engines, thus the cost can be greatly reduced.

2.4.2　Introduction to Full Digital Management System for MEA

Using electrical energy to operation is more efficient and more environmentally friendly than traditional aircraft systems. Conventional aircraft systems are powered by electrical energy, including generators, AC and DC loads, etc. This will lead to an increase in the number of electronic and electrical components, the increase of electric power required, and the management of electrical energy in the PSS becomes more important. There are many new technologies in the generation, storage and conversion system of MEA electrical energy, and there are still many related challenges.

Large aircraft combines flight control subsystem, avionics system and deicing subsystem. The generator is directly installed and connected to the engine, more easily to produce power supply interruption, so the management of aircraft PSS becomes very important.

Therefore, the energy management system of More-Electric Full Authorized Digital Electronic Control (MEFADEC) appeared.

1. The Energy Management System of More-Electric Full Authorized Digital Electronic Control

MEA engines benefit from engine electronic control technology. The Electronic Electrical Controller (EEC) system can make the engine run more efficiently, and in the operating environment of the control system, if the input sensor signal is evaluated, the useful input can be judged. With the gradual elimination of mechanical systems and the adoption of full power digital engine control with more-electric engines, maintenance intervals have been extended, resulting in lower engine maintenance costs and improved reliability.

Digital technology is the use of a computer as the core of the engine EEC to control the engine. The Full Authority Digital Controller (FADC) combines the throttle lever, strut and hybrid controller into one control unit to improve fuel economy of engine operation. The MEFADEC of the MEA will become a part of the engine, that is, the engine will become an MEA, reducing fuel consumption, resulting in higher efficiency and reliability. By adopting an MEA design strategy, some disadvantages of conventional FADEC, such as no manual control and speed limit, can be overcome.

The composition and structure of MEFADEC is shown in Fig. 2. 4. 2. With the increasing application of DC micro grid technology in aircraft control in aerospace field, the typical application in MEA is more-electric Starter/Generator(S/G). MEFADEC can design the steady-state operation performance of multi-channel high-power load. There are some typical loads in DC micro grid, such as voltage-controlled DC/DC converter, speed controllable generator of constant power load operation system and so on.

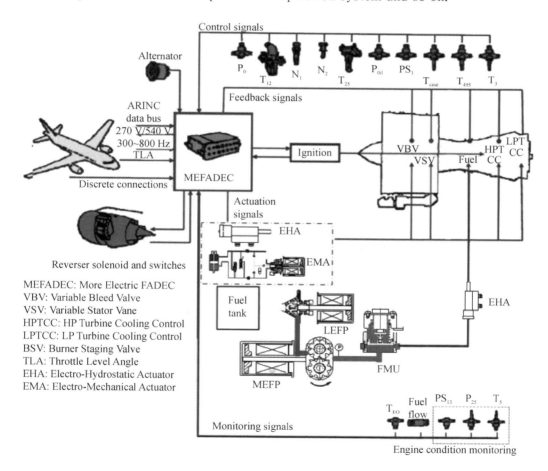

Fig. 2. 4. 2 Composition and Structure of MEFADEC

The MEFADEC receives signals from the ARINC data bus, and the power supply is 270 V/540 V high voltage DC power supply or 300—800 Hz variable frequency AC power supply, as well as throttle rod position signal. MEFADEC also receives engine condition monitoring signals. The main feedback signals from aircraft engines are Variable Bleed Valve (VBV), Variable Stator Vane (VSV), Variable Exhaust Valve (VEV), Burner Staging Valve (BSV), Throttle Level Angle (TLA), EHA, EMA, High Pressure Turbine Cooling Control (HPTCC) and Low Pressure Turbine Cooling Control (LPTCC).

The fuel in the system is supplied by the fuel tank, and is controlled by the Low-pressure Electric Fuel Pump (LEFP), Medium-pressure Electric Fuel Pump (MEFP), Fuel Measurement Unit (FMU) and EHA. The MEFADEC can also be integrated with the ignite of the engine and the engine.

2. The Control Law of MEFADEC

In the EEC control system, there are innerloop control and outer loop control. The inner ring is called the fuel flow cycle, which monitors the fuel flow of the engine fuel tank; the outer ring is the speed cycle, and when the pilot pulls the throttle lever, the engine opportunity enters different operations. There are a large number of control algorithms in the design of control system. For engine, it is generally about the control design of fuel supply and engine speed. The demand fuel flow value and the flow error are two important variables of fuel, and the state of the engine can be obtained from the control module of the engine model. In some complex cases, according to the fuel flow requirements, it is alternately the best choice of the control system under different engine conditions. Make a reasonable choice under all the circumstances that need to be considered by using the intelligent electrical control system.

The indispensable power electronic components such as EHA, motor and so on have the characteristics of high temperature and fault tolerance. The validity and correctness of the design conclusion are verified by simulation.

One of the key component of the MEA is the electric fuel pump, and its control system is extremely important. The MEA uses the motor to drive the fuel pump and controls the fuel flow by adjusting the speed of the fuel pump. The response speed and measurement accuracy of fuel flow depend on the response characteristics of electric fuel pump and the feedback accuracy of fuel flow. In order to improve the response ability of the motor, the direct drive motor is used to drive the fuel pump. The electric fuel pump controller receives the fuel control signal from the engine digital controller and tests the fuel flow through the feedback signal of the fuel flow sensor. According to the fuel flow difference between the control signal and the feedback signal, compared with the actual operation, the speed of the motor is adjusted to realize the fuel flow control.

As shown in Fig. 2.4.3, it is the control schematic diagram of the electric fuel pump. The digital engine controller in the figure consists of three parts, namely, motor controller, electric motor, high-pressure gear pump and so on. Fuel is transported to the high-pressure fuel pump through the low-pressure fuel pump, and then to the fuel nozzle through the high-pressure fuel pump. The flow sensor samples the high-pressure fuel and transmits the fuel signal to the motor controller. The motor control system adopts current loop, speed loop and position loop for three-loop control strategy, in which the current loop and speed loop adopt Proportional Integral Regulator (PIR), and the position loop adopts proportional P correction. The current feedback signal of Brushless DC Motor (BLDCM) can use bus current or phase current as feedback signal to form a current loop.

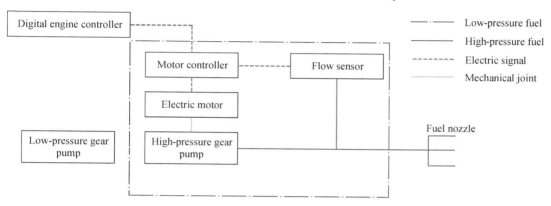

Fig. 2. 4. 3 Control schematic diagram of the electric fuel pump

The phase current is directly proportional to the torque of the motor, but the busbar current is different from each phase current. In order to ensure that the busbar current can accurately reflect the magnitude of the phase current, the sampling time is set to sample in the effective time interval of each Pulse Width Modulation (PWM) cycle.

2.4.3 Aviation Operation Requirements and Environmental Limitations

The hybrid power system of MEA is restricted by aviation operation requirements and environment, which is mainly reflected in the constraints of airborne equipment voltage and environmental conditions.

1. Airborne Voltage Level Limit

Under the propulsion of hybrid aircraft, the demand for electrical energy is greater, so the power supply voltage must be increased, and the voltage limit value may be reached 3 kV. With the development of hybrid aircraft technology, high-capacity electronic converters can be installed near engines or generators. The reliability of high voltage and high power electronic converter is the main challenge of architecture design. With the increase of altitude and the decrease of air pressure, the dielectric problem of insulating materials becomes another key factor. For example, for B787 and A350X aircraft, with the increasing power consumption on board, the grid voltage is designed as 230 V variable frequency AC voltage and 270 V/540 V high DC voltage.

2. Limitations of the Environmental Conditions

With the development of MEA and MEPS, the integration of power electronic devices faces challenges in three areas, namely, harsh environment challenges, challenges in volume, weight, and efficiency, etc. , and intelligent integration challenges.

(1) Harsh Environmental Challenges

Both the MEA and the MEPS must pay special attention to the low pressure environment. With the increase of voltage quota, the change of temperature will have a direct impact on power electronic components. For example, the heat generated inside a power semiconductor device can lead to an energy cycle effect. The thermal expansion coefficient of the materials used in the package is different, and the energy feedback and cycle produced by electromagnetic induction will produce thermal fatigue in the power assembly. Humidity and vibration will accelerate the decline in the performance of power components. High temperature and wide-range thermal cycle is a key challenge for electronic equipment. According to the current research status, the performance of capacitors, packaging materials, busbars and connectors is far from meeting the requirements.

(2) Challenges in Volume, Weight, Efficiency, etc.

The reduction of size and weight is one of the most important challenges in the field of aerospace. The application of power semiconductor with WBG can increase the switching frequency of power converters, thus reducing the size and weight of passive components such as filters and Electromagnetic Interference (EMI) components.

(3) Challenge of Intelligence and Integration

The intelligent and integrated design is adopted to modular, intelligent distribution of power converters to improve the system redundancy capability and reduce the total weight. Indeed, each power converter can be used for several loads. The converters and loads can be connected in a preset order, avoiding the use of a single power converter for the individual loads, increasing redundancy.

MEA and MEP are the next challenging goals for the aviation industry. Future aircraft will have to face emission reduction requirements through the addition of embedded power generation and distribution systems.

The cutting-edge solutions to improve the reliability of onboard power converters are: semiconductor devices, passive devices (magnetism and capacitors), power modules, gate drivers, power converter topology, EMI and filters. In addition, the cables in the harness, the power connectors that interconnect the power converter with the load, and the power supply interconnected with the converter will be studied. All these solutions require a high level of technology and need to be verified by experiments in order to maintain the attractiveness and safety of aerospace transportation.

2.5 Summary

This chapter mainly introduces the aircraft energy, that is, primary energy and secondary energy. This paper introduces the problems existing in the energy utilization of traditional aircraft, compares the differences between conventional engines and MEE, and the benefits brought by the unified use of electrical energy as secondary energy. The hybrid power system of MEA is briefly introduced.

2.6 Exercises

1. What is the primary energy source of an aircraft?
2. What are the secondary energy sources of traditional aircraft?
3. What are the benefits of the same secondary energy use of the aircraft?
4. What is the function of the full digital electronic controller for the MEA?
5. What is the value of high voltage DC and high voltage AC used in MEA?

Chapter 3　Power Supply System of MEA

With the increase of power consumption in airplane, there has been a structural evolution in aircraft Power Supply System (PSS), experienced low voltage DC PSS, Constant Frequency Power Supply System (CFPSS), Variable Frequency Power Supply System (VFPSS) and advanced High Voltage Power Supply System (HVPSS). This chapter mainly describes the characteristics of these systems.

3.1　The Evolution of Power Supply System

3.1.1　Low Voltage Direct Current Aircraft Power Supply System

Low Voltage Direct Current (LVDC) Aircraft PSS is suitable for small and medium aircraft. With the widespread development of more-electric technology in aircraft, the installed capacity of aircraft generators is increasing. If this system continues to be used, the heavy current will make the aircraft PSS very heavy. Therefore, 28 V low-voltage DC PSS is rarely used as the main PSS in large and medium-sized aircraft, but there are still a lot of avionic electronic equipment using 28 V DC PSS. Therefore, there exists the secondary power supply that converts the main power supply into 28 V DC power supply in MEA.

3.1.2　Constant Frequency AC Power Supply System

1. Constant Speed Frequency AC Power Supply System

Since the 1970s, the main power supply of the transport aircraft has always been three-phase 400 Hz/115 V AC. The conventional constant frequency is through the Constant Speed Device (CSD) to change the engine's mechanical energy output at varying speed into the mechanical energy at constant speed, and then drive the alternator. As shown in Fig. 3.1.1, it is a typical dual spool turbofan engine.

The best way to generate constant frequency is to use an Integrated Drive Generator (IDG), as shown in Fig. 3.1.2. IDG combines a CSD with a generator to make up a whole. Comparison with systems used separately with constant speed drives and generators, IDG is simple in structure, small in size, light in weight and easy to maintain.

As shown in Fig. 3.1.3, it is the schematic structure diagram of the constant speed and constant frequency AC PSS. There are two constant speed drives driven by the engine shaft, two constant speed drive Gen 1 and Gen 2, and two APU alternators driven by the APU on board. The main generator adopts a three-stage electric excitation synchronous generator.

The power from the main generator is 115 V/400 Hz, deliver electrical energy to the AC busbar, supply AC load on the aircraft. The main PPS uses an AC/DC converter(also

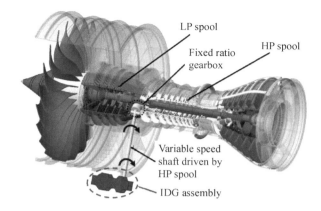

Fig. 3. 1. 1   A typical dual spool turbofan engine

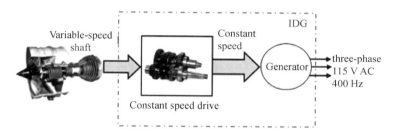

Fig. 3. 1. 2   Schematic diagram of IDG

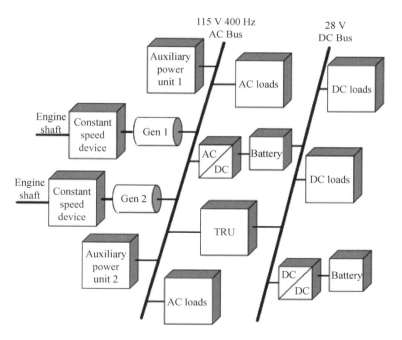

Fig. 3. 1. 3   Schematic structure diagram of the constant speed and constant frequency AC PSS

known as battery charger), charge the onboard battery. The main AC PSS can also transmit DC power to 28 V DC bus by using transformer rectifier unit(TRU). In addition to supply the DC load, this DC PSS can also charge the onboard storage battery.

In the CFPSS, AC/DC converters and TRU are the primary power converters, batteries serve as an emergency source of power. The AC/DC inverter and the bi-directional DC/DC converter connect the storage battery voltage with the AC busbar and the DC busbar respectively.

Limited by the engine space, Other AC or DC generators cannot be installed. Long cables are also needed to connect the generator to the busbar. A TRU is introduced to convert AC voltage into DC voltage, because it can be installed near the DC busbar, this will reduce the weight and volume of the aircraft PSS to a relatively low level. TRU can also reduce harmonic currents, the Total Harmonic Distortion (THD) is reduced more.

Although CSCF aircraft PSS is still widely used in various military and civil aircraft at present, and has been greatly improved after decades of development, it has always had defects of varying degrees in reliability, maintainability, weight, cost and survive ability of battle loss.

2. Variable Speed Constant Frequency AC PSS

Over the past 20 years, alternative methods of generating constant frequency power have been tested and used on a number of civil aircraft. This is an attempt to convert a variable frequency power source generated by a generator directly driven by the gearbox of a variable speed engine accessory into a constant frequency power source by means of an electronic variable frequency device. This is called Variable Speed Constant Frequency (VSCF) technology, as shown in Fig. 3. 1. 4.

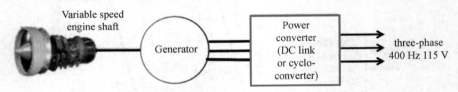

Fig. 3. 1. 4　Schematic diagram of VSCF

There are usually two methods for power converts level, as shown in Fig. 3. 1. 5, where Fig. 3. 1. 5(a) adopts AC/DC, DC link and DC/AC converter, and an AC/AC power conversion method in Fig. 3. 1. 5(b).

Compared to CSCF aircraft PSS, VSCF aircraft PSS system has the advantages of good electrical performance, high efficiency, high reliability and low maintenance cost. Therefore, it was once highly valued. However, the high-power VSCF PSS is mainly limited by the power devices at that time, and the technology can not meet the expected reliability requirements. Later, after exploration and research, a three-stage brushless alternator system was developed and applied, therefore, the CF Aircraft PSS interview with constant

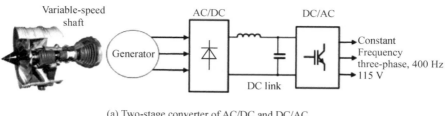

(a) Two-stage converter of AC/DC and DC/AC

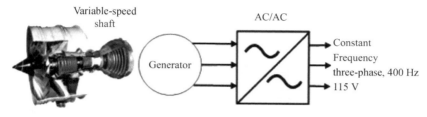

(b) An AC/AC power converter

Fig. 3. 1. 5 VSCF power supply system using power converter

speed transmission device was canceled.

3. Three-stage Brushless AC Power Generation System

Fig. 3. 1. 6 shows the schematic which is normally driven by an accessory gearbox on a jet engine. The generator is a three-stage permanent magnet synchronous generator. Generator Controller Unit (GCU) regulates the excitation field of the main generator, thus further adjusts the generator output voltage. In Fig. 3. 1. 6, the first stage is a permanent magnet exciter stage, the second stage is an Alternator Current Exciter(ACE), and the third stage is the main generator. Permanent magnet and rotary rectifier are installed on the rotor, which is one of the key components of three-stage brushless AC generator to realize brushless.

The output current and voltage of the second stage ACE should meet the excitation requirements of the main generator. The main generator maintains the rated output voltage within the no-load and rated load range. Aircraft alternators are also required to be 50% overloaded within 5 minutes and 100% overloaded within 5 seconds, and the output voltage of the generator shall still be the rated value. The generator is also required to have a steady state short circuit current output of 300% rating within 5 seconds. Therefore, the aircraft alternator's excitation current should operate from its minimum value (no load) to its maximum value (100% load and 300% overload for long term operation).

The operating environment of the alternator is within the range $-60-70$ ℃. In case of oil cooling, the temperature of the cooling oil varies between 120 ℃ and 140 ℃. The thermal resistance of the generator excitation winding is almost twice the cold resistance. In such a wide range, the exciter cannot operate at saturation, but keep the output voltage stable. The three-stage brushless AC generator has been used in many models and is still in use.

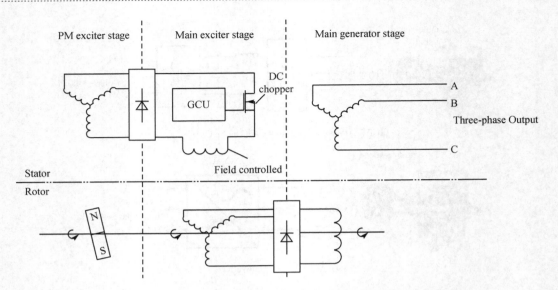

Fig. 3. 1. 6　Schematic diagram of a three-stage brushless AC generator

3.1.3　Variable Frequency PSS

As early as during World War Ⅱ, some high-powered aircraft used variable frequency AC power supplies. Its frequency range is narrow, modern turboprop aircraft and helicopters also use variable-frequency AC power supplies. With the increase of installed capacity and the change of power supply requirements for electrical equipment, this narrow frequency conversion AC power supply cannot meet the requirements.

As shown in Fig. 3. 1. 7, it is the structure diagram of the variable frequency generator directly driven by the engine. Due to its output frequency depending on the output speed of the engine, especially most aircraft use turbojet engine or turbofan engine. Engine speed variation range is large, so this frequency conversion AC power system is called wide frequency conversion AC power system, it has the disadvantage of large frequency change, its development has been limited.

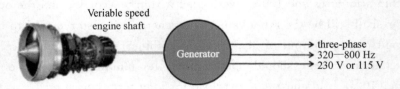

Fig. 3. 1. 7　A variable frequency generator directly driven by the engine

With the development of power electronics technology and its application in aircraft, the variable frequency AC power system is easier to form the variable frequency AC start power generation system and has been applied in the large civil aircraft, such as the B787 and A380. The main types of aircraft using variable-frequency power are as listed in Tab. 3. 1. 1.

Because the alternator is driven directly by the motor transmission box of the engine accessory, its speed changes with the speed of the engine, and the frequency variation range is large, generally about 2:1.

Tab. 3.1.1   Civil aircraft with VFPSS

SN	Name	Country	Production	Capacity/(kV · A)	Range/Hz
1	Belfast	UK	1996	8×50	334—485
2	Global express	CAN	1994	4×40	324—596
3	Xinzhou – 60	CN	1997	2×20	325—528
4	B787	USA	2009	4×250	360—800
5	A380	EUR	2008	4×150	360—800

Fig. 3.1.8 shows the AC PSS of a variable frequency aircraft. Since the generator is connected directly to the engine shaft, when the voltage of main AC bus is 115 V, the operating frequency is 360—800 Hz. A variety of converters are used to transform VF AC power supply into a variety of power supply voltages, for example, 115 V/400 Hz AC and 270 V DC.

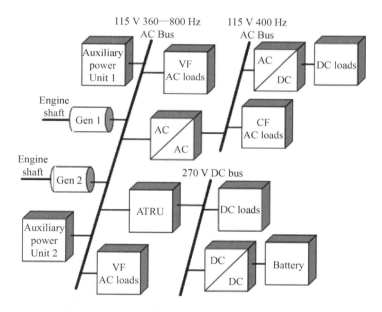

Fig. 3.1.8   AC PSS of a variable frequency aircraft

It is worth noting that the frequency of the main AC Bus varies from 360 Hz to 800 Hz. The frequency is proportional to the speed of the engine shaft. Through the use of VF PSS, the bulky and inefficient Constant Speed Device(CSD) was eliminated, so that the aircraft power system PSS can obtain better performance.

As the frequency of the main AC busbar is variable, the variable frequency AC power supply needs to be converted into aircraft CF PSS for the electrical load that requires constant frequency power supply. As shown in Fig. 3.1.9, a back-to-back converter structure is used to generate 115 V/400 Hz constant frequency AC to supply load. These converters must be carefully designed to meet volume, weight, and harmonic requirements as described in the "Environmental Conditions and Test Procedures for Airborne Equipment" DO – 160 standard or the military standard MIL – 704.

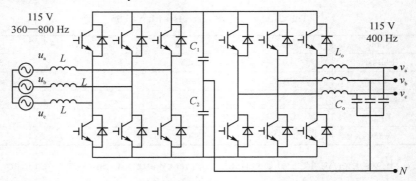

Fig. 3. 1. 9   Back-to-back converter structure

As shown in Fig. 3. 1. 9, two PWM converters are used in the back-to-back converter. Among them, the first-level high power factor rectifier adopts D-Q decoupling strategy to control active and reactive current, which can make the power factor close to 1 and enable the current harmonics to meet higher standard requirements; The second stage uses a sinusoidal SPWM inverter to transform the DC voltage of the first stage into a constant frequency of 115 V/400 Hz AC low voltage. The circuit adopts voltage and current double loop control scheme to make the bus voltage and frequency stable. The DC bus voltage was chosen for 270 V because it can be obtained directly by rectifying the 115 V AC voltage. DC/DC converters are used to obtain a variety of DC voltage values.

Auto-transformers are much smaller in size and weight than isolation transformers at the same power level. Therefore, the function of transformer rectifier TRU can be replaced by Auto Transformer Rectifier Unit (ATRU). Fig. 3. 1. 10 shows the topological structure

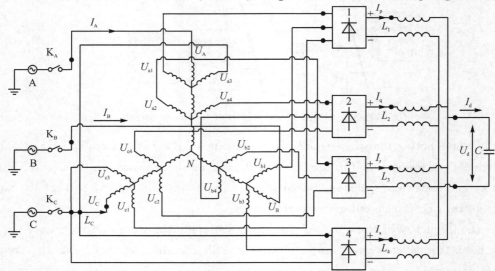

Fig. 3. 1. 10   Topological structure of 24-pulse ATRU

of 24-pulse ATRU.

In addition, bi-directional DC/DC converters are used in aircraft with variable frequency AC power to charge and discharge storage batteries as emergency power sources.

3.1.4 Future Aircraft PSS

In the recent PSS design of aircraft power system, as the energy demand of hydraulic, pneumatic and mechanical systems is gradually replaced by electrical energy, the electrical energy demand has an obvious trend of growth. However, the feeder cable current in traditional 115 V or 28 V PSS will increase proportionately to the power rating, which will certainly lead to higher power losses and cable weight.

Therefore, increasing the voltage value of future aircraft is clearly better than continuing to use 115 V AC and 28 V DC low voltage power systems to reduce the current in the feeder cables and further reduce the weight of the cable, resulting in lower losses and higher efficiency.

1. High Voltage AC (HVAC) Aircraft PSS

Recent aircraft have adopted a high voltage AC power system, HVAC, or 230 V AC, at a frequency of 360—800 Hz. For example, the B787 has reduced transmission loss and converter weight by 50.7% and 42.5%, compared with conventional 115 V PSS.

Fig. 3.1.11 shows the HVAC aircraft PSS, the system has two main generators driven directly by the engine and two auxiliary power units, they directly output 230 V AC variable frequency AC power supply for variable frequency AC load. The main AC power supply is converted to 270 V DC by the ATRU, and the Auto Transformer Unit (ATU) is used to generate 115 V variable frequency AC power supply. Use DC power supply generated by DC/DC converter to charge storage battery.

Fig. 3.1.12 shows the topology of Buck Boost Converter Unit (BBCU). Firstly, 270 HVDC is transformed by full-bridge circuit, step-down by high frequency transformer, and then 28 V DC power supply is generated by diode bridge rectifier circuit and filtering.

2. HVDC Aircraft PSS

HVDC aircraft PSS have several possible configurations, i. e. :

① ±270 V DC, ground wire with midpoint.

② 270 V DC, direct grounding.

③ ±135 V DC, ground wire with midpoint.

④ ±135 V DC, ground without midpoint.

Studies have proved that in most cases, PSS can reduce the weight of the high voltage DC PSS, and the weight reduction can be from 4% to 28%. Fig. 3.1.13 shows a 270 V HVDC aircraft PSS. Fuel cell will replace turbo-powered APUs, which are usually less than

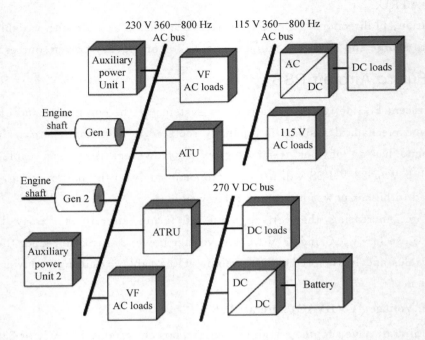

Fig. 3. 1. 11   HVAC aircraft PSS

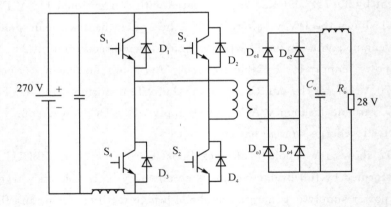

Fig. 3. 1. 12   BBCU topology diagram

20% efficient together with noise and gas emissions.

Some topology have been reported to integrate fuel cell systems into aircraft power systems. Fig. 3. 1. 14 shows the parallel architecture of fuel cell systems.

3. Challenges for Future Aircraft

Currently, electrical equipment does not provide better reliability, but can increase availability because subsystems can be isolated during failures. Although failures can be predicted by modeling the behavior of the system in conjunction with fault detection algorithms, a better approach is to develop highly reliable, high-power density power

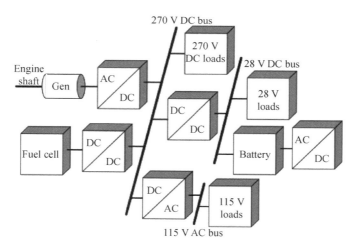

Fig. 3. 1. 13   270 V HVDC aircraft PSS

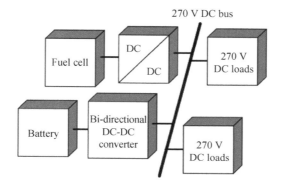

Fig. 3. 1. 14   Parallel architecture of fuel cell systems

electronics devices. For example, semiconductor devices based on silicon carbide SiC show great promise because it can significantly reduce switching losses, high temperature tolerance and fast switching capability.

At the frequency converter level, PWM power converter has a very complex control strategy, which may lead to the instability of the whole distribution network. Some studies show that the converter is guaranteed to work in the stable region. It is worth mentioning that sharing power electronic converters is an effective way to save quality.

At the system level, the integrated optimization design will become the main research field in the next ten years because the traditional testing method cannot guarantee the optimal operation point of the system. The integrated optimization design will balance system mass, efficiency, thermal stability and power quality through an advanced algorithm.

3.2   Introduction to Hybrid Excitation Generator

3.2.1   Mixed Excitation Brushless Aircraft Generator

1. Introduction

In order to improve the reliability and efficiency of conventional aircraft and reduce fuel consumption, it is necessary to develop MEA, which is based on the use of electricity to drive aircraft subsystems originally driven by a combination of electric, pneumatic, mechanical power transmission systems, etc. With the development of MEA, the constant speed drive device has been eliminated. The research on Variable Frequency AC (VFAC) power generation system has become a focus of attention.

The three-stage brushless generator consists of a permanent magnet generator, a main synchronous exciter and a synchronous generator, which is attractive for their voltage control capability and high power density. The main synchronous generator's excitation current is generated by a combination of the exciter and the rotary rectifier. But because of using the rotary rectifier, the axial dimension is long, and the reliability is reduced. Hybrid Excitation Synchronous Machines (HESM) have attracted much attention because the flux density in the air gap can be easily adjusted by adjusting the current in the excitation winding. Due to the advantages of HESM in high power density, brushless and simple structure, the combination of Permanent Magnet Synchronous Machine (PMSM) and Wound Rotor Synchronous Machine (WRSM) is particularly suitable for VFAC power generation system. At the same time, HESM is not as reliable as rotary rectifier. The brushless aircraft generator with mixed excitation is the most potential alternative generator for variable frequency AC power generation system.

2. Mixed Excitation Generator

Fig. 3.2.1 shows the structure diagram of the hybrid excitation generator. Fig. 3.2.1 (a) shows the structure diagram of the HESM with a rotor with parallel excitation. The stationary has 54 slots for laying armature winding. The rotor has rotor cores forming N poles and S poles, 12 permanent magnets PM embedded in the rotor slots, and the rotor has excitation winding and matching rectifier Bridges. It can be seen that HESM is evolved from the traditional tangential Permanent Magnet Synchronous Motor (PMSM). The N-pole rotor core extends along the axis to form a circular magnetic field, while the S-pole rotor core shrinks inward at the end and then extends along the axis to form another circular magnetic field.

Fig. 3.2.1 (b) shows the rotor shape diagram of HESM hybrid excitation motor. In order to increase magnetic flux of permanent magnet and reduce flux leakage, the method of increasing axial length of rotor and combining excitation winding is adopted to increase flux.

At the same time, brushless excitation is realized because of the auxiliary air gap between rotor core extension and magnetic bridge. This type of motor produces a large electromagnetic force due to the special structure of magnetic bridge and rotor core extension.

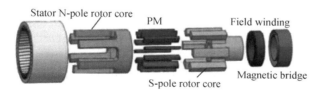

(a) HESM with parallel magnetically excited rotor

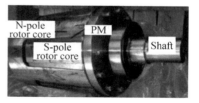

(b) Rotor shape diagram of HESM hybrid excitation motor

Fig. 3. 2. 1   Structure diagram of the hybrid excitation generator

3.2.2 Overview of B787 More-Electric Generator

With many novel MEA features, the B787 has taken a big step towards AEA. This is mainly reflected in the fact that all systems are electrically driven. Except for the anti-icing of engine air intake, engine air intake is basically canceled. Although hydraulic actuators are also applied, most of their power is supplied by electric power supply.

1. Variable Frequency AC Starter/Generator for MEA

The power weight ratio of more-generator is greatly improved. Due to the limited number of engines carried by the aircraft, the number of generators carried by the aircraft is also limited. Therefore, the installed capacity of a single generator is a very important technical index, which relates to the power supply capacity of the whole machine.

The bypass ratio of turbofan engine is increasing continuously, and has reached 10:1 at present. The increase of bypass ratio leads to the increase of the loss caused by air entraining, and the elimination of air entraining has become an important trend of engine development. The B787's engines extract only a small amount of air behind the fans to prevent icing in the engine's intake ports, saving 20% fuel for the plane (compared with conventional aircraft).

The structure of more-generator is quite different from that of traditional generator. The structure of generator with constant speed transmission device is very complex, while the structure of variable frequency starter/generator is much simpler, and its reliability and power density are greatly improved. The Constant Speed Constant Frequency (CSCF) power supply system must be abandoned, because CSCF system must have a CSD and a GEN power transformation and the power generation efficiency is only about 70%. If you want to generate electric power of 500 kW, there is a loss of nearly 240 kW. This increases the fuel consumption, increases the heat dissipation equipment, and greatly reduces the reliability. If the variable frequency alternator is used, the efficiency can reach 90%. It can produce power of 500 kW, and the loss is only 55. 6 kW. It greatly reduces the loss and the volume weight, improving the reliability.

Take B787 as an example, there are two engines with four generators respectively, and there are two APU generators, which are directly driven by the main engine and the APU starting engine driven by the auxiliary power device.

Fig. 3. 2. 2 shows the installation of variable frequency starter/generator on the engine of B787. The engine is connected to the generator shaft through the accessory gearbox. The power of VFSG is 250 kV · A, the operating frequency is 360—800 Hz, the weight is 203 lbs(1 lbs=0. 453 59 kg), and the Mean Time Between Failure (MTBF) is 30 000 flight hours. The specific parameters are listed in Tab. 3. 2. 1.

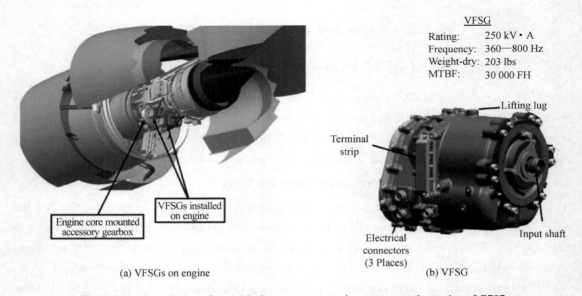

(a) VFSGs on engine   (b) VFSG

Fig. 3. 2. 2   Installation of variable frequency starter/generator on the engine of B787

Tab. 3. 2. 1   B787 starter/generator technical data

SN	Technical data	Value	SN	Technical data	Value
1	Rated voltage/V	230(P)/400(L)	8	Speed/(r · min^{-1})	7 200~16 000
2	Rated current/A	361	9	Start torque/(N · m)	⩾400
3	Rated frequency/Hz	360—800	10	Cooling way	Spray oil cooling
4	Power rating/(kV · A)	250	11	Efficiency/%	90
5	Overload capacity	120%, 5 min 175%, 5 s	12	MTBF/FH	30 000
6	Power factor	0. 8—1. 0	13	Net weight/(kg · lbs^{-1})	91/203
7	Pole-pairs	3	14	—	—

As shown in Fig. 3. 2. 3, it is the cross-section of VFSG. It is composed of three-stage structure composed of permanent magnet generator (pilot exciter), exciter and main generator. Without the traditional constant speed device, the structure is greatly simplified, the power density is greatly improved, small volume, light weight, and the average fault— free time is long.

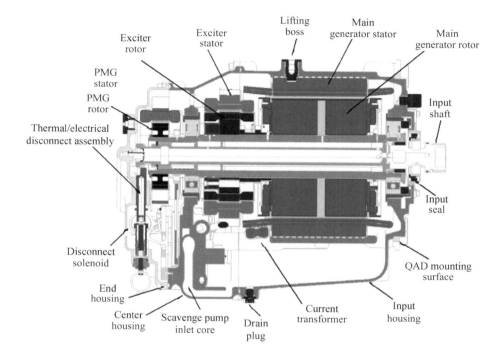

Fig. 3. 2. 3   Shows a section view of a VFSG

If variable frequency AC generator in high speed work, suddenly unload high-power electrical equipment, voltage regulator of delay will lead to a dramatic increase in generator voltage and do harm to electrical equipment. The instantaneous voltage limiting circuit is used to quickly reduce the over voltage of the generator. In practice, it is used to slow down the unloading speed of the generator to limit the instantaneous over voltage of the generator.

2. The APU Starter/Generator

The APU generator of the AC power system also adopts the variable frequency generator. Fig. 3. 2. 4 shows the appearance of the APU Starter/Generator (ASG). Each APU has two Starter Generators, and each generator has a rated power of 225 kV · A, a frequency of 360—440 Hz, a net weight of 122. 7 lbs, and the mean time between failure is 40 000 hours.

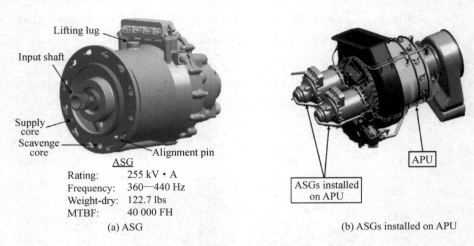

ASG
Rating:　　　255 kV · A
Frequency:　 360—440 Hz
Weight-dry:　122.7 lbs
MTBF:　　　 40 000 FH

(a) ASG　　　　　　　　　　　　　　(b) ASGs installed on APU

Fig. 3. 2. 4　Appearance of the APU starter/generator

3.3　Variable Frequency Power Supply System of MEA

The PSS of MEA is the same as that of conventional aircraft, which is composed of main power supply, auxiliary power supply, emergency power supply, secondary power supply and ground power supply interface.

3.3.1　Main Variable Frequency AC Power Supply System

Due to the limited number of engines that aircraft can equip, the number of generators is limited. Due to the sharp increase in electricity consumption, the installed capacity of a single generator is increasingly required. Due to the rapid development of secondary power conversion technology, the variable frequency AC power system appeared. In this way, the power quality problems caused by load and engine speed changes are solved by the secondary power converter.

Variable frequency AC power system has the advantages of simple structure, high energy conversion efficiency and high power density. The variable frequency AC power system is composed of an alternator and a controller. The system has only one transformation process. The alternator is driven directly by the motor transmission box of the engine accessory. There is no constant speed transmission device (constant speed constant frequency system) and secondary conversion device (variable speed constant frequency system), easy to constitute a starting dynamic power generation system. Therefore, in terms of PSS itself without considering distribution system, electrical equipment and engine starting factors, in a variety of PSS scheme, variable frequency AC power generation system has the simplest structure, the highest reliability, the highest efficiency, the lowest cost and other advantages, and has a small weight and volume.

1. The Power and Density of the Power Supply Increases Significantly

Fig. 3. 3. 1 shows the growth of aircraft power consumption, in which the power consumption of A380 reaches 800 kV · A and that of B787 reaches 1 450 kV · A.

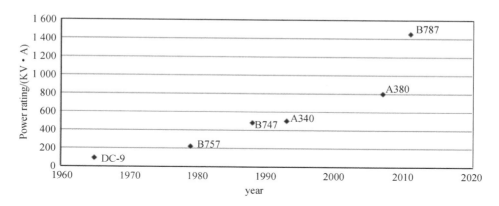

Fig. 3. 3. 1   An increase in aircraft power consumption

The power of MEA is the sum of the power of conventional aircraft power supply, hydraulic power source and pneumatic power source. The hydraulic power of modern high-performance aircraft is much larger than the power supply. The power capacity of MEA increases sharply, and the power supply tends to increase with the development of aircraft.

For example, F – 16A/B's main power supply is 40 kW, F – 16C/D's is 60 kW, and JSF35's main power supply is 250 kW/270 V switched reluctance start dynamo system, and 500 kW start dynamo system is being developed abroad.

2. The Quality of Power Supply Should be Further Improved

Aircraft electronic equipment on the power supply voltage accuracy, stability, interrupt time has strict requirements, supply power outage for a short period of time can make electronic equipment cannot work normally, which affects the other aircraft systems and threatening flight safety. Power conversion is the inevitable requirement of the rational use of the power supply, power conversion process, especially to protect the operating process of the power outage is not avoidable, uninterruptible power supply can only be achieved when the power supply is operating normally, which poses a challenge to the design of all-electric and MEA. So the technology of uninterrupted power supply has made some progress through many years of research. The technology of fast detection, solid state device, super capacitor and high voltage DC power supply is helpful to solve the technical problems in uninterrupted power supply.

(1) The Microprocessor Increases the Requirement of Uninterrupted Power Supply

With the rapid development of aviation technology, microprocessors are widely used in airborne electronic equipment. For example, there are 51 systems using CPUs on Boeing 767 aircraft. In order to be safe and reliable, many systems are equipped with 2 to 3 redundant systems, the entire aircraft uses more than 170 CPUs, the use of new aircraft microprocessor is larger. When these devices are operating, some data may be lost and cannot operate properly if power is suddenly cut off. Despite the addition of auxiliary circuits to the PPS and electronic equipment to reduce the interference of power interruption to the normal operation of the equipment. However, power interruption still occur from time to time, which has become a hidden danger of flight safety. Therefore, it is imperative to adopt the uninterrupted technology in the normal power conversion, especially in the abnormal operation of the power supply system.

(2) Super Capacitor Energy Storage Advantages

Super capacitor is a special capacitor for energy storage developed in recent years. The energy volume density is hundreds of times higher than that of traditional electrolytic capacitor, and the leakage current is nearly a thousand times smaller.

The discharge power ratio of super capacitors is nearly ten times higher than that of the battery, which can quickly store and discharge, suitable for a wide temperature range, long life, especially suitable for the need of instantaneous high power occasions.

Super capacitors do not need any maintenance, the life of more than ten years, is an ideal capacitor.

Both in DC and in the AC system, when the fault generator out of the grid, the aviation battery will bear larger impact load. Large current discharge will have serious influence on the life of the battery. Super capacitor can overcome these disadvantages of battery. It is suitable for a short time grid large current power supply, and it can also be used as engine air start power supply.

3. Low Power Loss and Heat and High Efficiency is Required for Main Power Supply

This requirement rules out the use of constant speed CF aircraft PSS and variable speed constant frequency power sources on aircraft. Because these two power sources have two energy transformations in the process of mechanical energy conversion into 400 Hz AC, the efficiency of constant speed constant frequency power supply is less than 72%, and the efficiency of variable speed constant frequency power supply is about 80%. Switch resistance starter/generator have promising applications, and is said to have an efficiency of 85% with a capacity of 250 kW(speed ranges from 13 000 to 25 000 r/min).

Most aviation power sources at home and abroad believe that 270 V HVDC power supply is the best power supply for MEA. The main power generator of 270 V HVDC power supply can be selected from three types: rotary rectifier type brushless synchronous generator, switched reluctance generator and electromagnetic double-pole generator.

Rotary rectifier type brushless synchronous generator is the only suitable motor for constant speed CF aircraft PSS. It has been developed greatly in the past 50 years, and its power to weight ratio is quite high. Most of the early aircraft variable speed constant frequency power supply also adopts this motor. At the end of the 20th century, this concept had a breakthrough: variable speed constant frequency power supply began to consider using switched reluctance or asynchronous generator. The rotating rectifier makes brushless synchronous generator with complex structure and high failure rate. The maximum speed is also limited, which limits the installed capacity of single machine.

The rotor structure of switched reluctance motor and squirrel cage asynchronous motor is simple and does not need exciter. Both have the function of starting power generation. The main deficiency is that power electronic converter must work with power generation.

At present, the proposed synchronous motor of two sets of isolated motor winding, one for excitation during power generation and the other one for output power can reduce the power of controllable electronic converter and improve the performance of the generator, but the starting efficiency of this structure is low.

The dual salient generator is similar to the brushless synchronous generator with rotary rectifier, the output is connected to the rectifier to obtain direct current, but its rotor structure is as simple as a switched reluctance motor. Starting and suspension of the main power motor is another major requirement for the generator used in all-electric engines. Switched reluctance motor, asynchronous motor and double salient pole motor have good reversible performance, which are easy to realize motor operation, can start the engine. The existing structure scheme of all-electric engine adopts active control radial and axial electromagnetic bearings, plus mechanical backup bearings. The appearance of bearingless motor provides a new structure scheme for all-electric engine, which helps to further improve the overall performance of the engine.

3.3.2 Characteristics of APU Power Supply of Auxiliary Power Unit

APU power supply is an indispensable power supply for modern aircraft. The APU is a combination of a small jet engine and a starter/generator. The combined power unit is the development direction of the auxiliary power unit, which has three functions: the starting function of the main generator, functions of auxiliary power units and functions as emergency power units. The latter requires a shorter start-up time for the combined power unit to ensure rapid input of emergency power in the event of a main power failure.

In the early space shuttles developed by the United States, hydrogen and oxygen fuel cell was the main power source of the aircraft. The development of fuel cell in the last 20 years has paved the way for their use as auxiliary power sources in aircraft. Fuel cell has high energy conversion efficiency and no pollution. It can get 270 V or 28 V direct current when combined with DC converter, and 115 V/200 V 400 Hz AC when combined with inverter.

MEA/AEA will no longer require a lot of ground support equipment as long as the ground power is supplied, but the power and quality of the ground power will be significantly improved.

There are many types of emergency power sources used by foreign aircraft, including Ram Air Turbine (RAT) driven generator, hydraulic motor driven generator, fuel emergency power generator and battery. The first two are mostly used for long-range aircraft, which have unlimited power supply time, while the last two have limited energy storage, thus shorting emergency power supply time. In the aircraft using low-voltage DC power supply, due to its simple structure, most batteries are used as emergency power supply and backup power supply. The combination of permanent magnet generator and converter driven directly by the aero engine is a newly developed emergency power supply, which has a wider operating speed range than the existing main power generator.

As shown in Fig. 3.3.2, RAT emergency PSS can provide emergency AC power for the flight control system by pounding the turbine blades with head-on airflow. It is a disposable device. When the main power supply is normal, it is installed on the abdomen of the aircraft and falls off when all the main power supply fails.

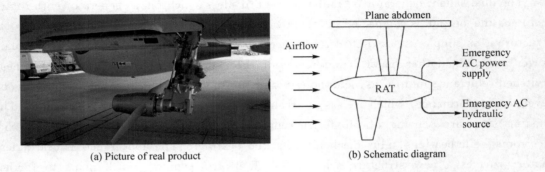

(a) Picture of real product   (b) Schematic diagram

Fig. 3.3.2   RAT emergency power supply(B787)

3.3.3   Characteristics of Secondary Power Supply

A secondary power source is indispensable on an aircraft. It converts the electrical power generated from the main power source into the type of power needed for certain electrical equipment. If the internal power sources of electronic equipment are included in the secondary power supply range, the number of secondary power sources on modern aircraft is considerable. As early as 10 years ago, foreign MEA developed 27 V DC/28 V DC, power of 5.6 kW DC converter efficiency of 90%, weight of 4.04 kg. Another 270 V DC/115 V AC, 400 Hz, capacity of 8 kV · A static converter efficiency up to 87%, weight of 8.3 kg.

270 V high voltage DC power supply is a typical distributed PSS. Secondary power supply is one of the keys to achieve redundancy fault tolerance and uninterrupted power supply. The modularization technology, parallel technology, fault detection, diagnosis and protection technology of secondary power supply are necessary for MEA/AEA secondary power supply.

3.3.4   Worth Research Problems

The development of variable frequency AC power has broken through the limitation of CF aircraft PSS and significantly improved the energy conversion efficiency and power generation, but there are also deficiencies of variable frequency AC power and load in adaptability. For example,

① The feeder of variable frequency AC power supply increases reactance and line voltage drop at high frequency, and the three-phase voltage becomes worse symmetrically under asymmetric load.

② Variable frequency AC power supply cannot operate in parallel.

Because the variable frequency AC power supply cannot operate in parallel, the conversion between the power supplies will lead to the interruption of power supply for some electrical equipment, which is absolutely not allowed for modern high-performance aircraft. On an aircraft using a variable frequency AC main power supply, only the secondary 28 V DC busbar can achieve uninterrupted power supply.

③ The problem of sudden unloading of variable frequency alternator at high speed.

When the variable frequency alternator rotates at high speed and suddenly discharges the high-power load, the voltage regulator has adjustment delay, which will cause the output voltage of the generator to rise sharply and endanger the electric equipment. Over voltage absorption circuit is often used to quickly absorb the over voltage of the generator, in fact, to slow down the unloading speed of the generator to limit the instantaneous over voltage of the engine.

④ The cable is too heavy due to the large transmission current.

Due to the large increase of installed capacity and the increase of the highest frequency of variable frequency, considering the skin effect, copper cable will make the cable too heavy. Because the skin depth of the wire is proportional to the square root of the resistivity, the DC resistivity of the aluminum wire is larger than that of the copper, so the skin depth of the aluminum wire is larger than that of the copper wire. In addition, the density of the aluminum wire is smaller than that of copper, so the aluminum wire can greatly reduce the weight of the wire. The A380, for example, has lost 400 kg of weight. The structural material of the A380 has been changed from aluminium to composite material, reducing its structural weight by 22%. Then, due to the use of aluminum wire, the line voltage drop will increase, and the resulting voltage fluctuation will be solved by the secondary power supply of the later stage of the electrical equipment.

⑤ Constant power load characteristics affect the stability of the power system.

A large number of DC/DC, DC/AC and AC/AC converters are presented with constant power load characteristics and negative impedance characteristics, which will cause the instability of the power system, and the current surge caused by the capacitive load. These require sophisticated design to be circumvented and can also cause circuit complexity.

3.3.5 Main Technical Data of B787 Power Supply

Tab. 3.3.1 lists the statistics of B787 power supply and its main technical performance. It can be seen from Tab. 3.3.1 that the total power generated by the main generator, APU generator and RAT generator reaches 1 460 kV · A.

Tab. 3. 3. 1   **B787 power supply technical data and its main technical performance**

SN	Name	Number	Technical date and main technical performance
1	VFAC/S/G	4	250 kV · A, 360—800 Hz, 230/400 V AC, oil cooling, 92 kg(dry weight). MTBF:30 000 FH
2	APU/S/G	2	225 kV · A, 360—440 Hz, 230/400 V AC, oil cooling, 55. 6 kg (dry weight). MTBF: 40 000 FH
3	RATG	1	10 kV · A, VF, 230/400 V AC
4	Ground power interface	4	90 kV · A, 400 Hz,115/200V AC
5	ATU	2	90 kV · A, 360—800 Hz,115/230 V AC bi-direction
6	ATRU	4	150 kV · A, 360—800 Hz,230 V AC/±270 V DC water-cooling
7	TRU	4	Input: 230 V AC 360—800 Hz; Output: 28 V DC 240 A
8	BPEU	4	2. 5 kW, 28 V DC/±130 V DC
9	Main BAT	1	48 A · h, lithium-ion battery, 24 V DC
10	APU BAT	1	48 A · h, lithium-ion battery, 24 V DC
11	BC	2	Boost converter
12	APU start converter	1	Boost converter

Fig. 3. 3. 3 shows the general situation of the PSS of B787. In addition to the power supply of B787, there are also five busbars of different voltage levels, namely, 230 V AC busbar, 115 V AC busbar, 28 V AC busbar, high voltage DC back flow bar and DC wheel brake busbar (not marked in the figure).

The DC PSS of B787 consists of four Transformer Rectifier Units(TRU), four Auto Transformer Rectifier Units (ATRUs), four Electric-Brake Power Supply Units (E-BPSUs), main battery, APU battery, flight control DC power supply and other subsystems. The main battery and APU battery are 48 A · h lithium-ion batteries.

1. ±270 V DC Bus

As shown in Figure 3. 3. 4, it is the B787 270V DC power supply schematic diagram, B787 has four 270 V DC busbars, which are as follows: L1±270 V DC, L2±270 V DC, R1±270 V DC and R2± 270 V DC. Four 270 V DC busbars are converted through four 230 V AC bars under the control of the respective Auto Transformer Rectifier Unit Controller (ATRUC), through the auto transformer rectifier unit, most of them are used for the power supply of the motor controller, adjusting the speed of the motor. There are nearly 12 kinds of loads, each power of 150 kW, using liquid cooling mode.

270 V DC generator usually has two forms, one is switch magnetic resistance starter/generator; the other is winding generator and rectifier. The biggest advantage of the 270 V DC is in the transmission of electricity, which is summarized as follows:

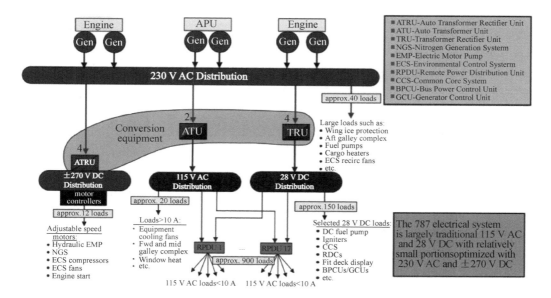

Fig. 3.3.3   PSS of B787

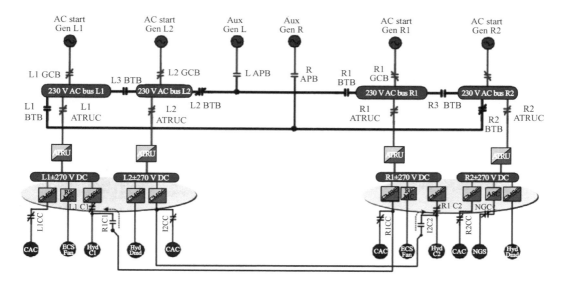

Fig. 3.3.4   The 270 V DC power supply of B787

(1) The Number of Wires is Reduced

AC transmission requires a three-phase, four-wire structure, while DC transmission requires only two wires, greatly reducing the size and weight of the cable.

(2) No Restriction Caused by Skin Effect

In addition, AC effect, namely skin effect, should be considered in the transmission of AC power, so that the diameter of the selected wire must be less than twice the skin depth. For large current transmission, multiple strands must be selected, with the same volume and weight increase. Direct current has no skin effect.

(3) High Security of Power Supply

270 V DC has the advantages of simple structure, high energy conversion efficiency, high power density, easy to realize uninterrupted power supply and safe use.

(4) High Power Efficiency

The power supply efficiency of the 270 V DC can reach more than 85%, while the power efficiency of the constant speed and constant frequency AC power supply is about 68%. Improving efficiency brings great benefits to reducing fuel costs or flight range.

However, there are still some technical and manufacturing problems with 270 V DC power supply, the main problem is: The control switch of the power supply system needs to be completely improved, from the original contactor, relay and circuit breaker to the Solid State Power Controller(SSPC) using solid state electronic technology, computer control technology and software technology, etc.

In military aircraft, the 270 V DC has been used in the F – 22 and F – 35 fighters, with the F – 22 using a 65 kW 270 V DC and the F – 35 using a 250 kW, 270 V DC starter/generator. The American F – 14A fighter aircraft, S – 3A and P – 3C anti-submarine aircraft locally use 270 V DC technology.

2. 120 V Electric Brake Power DC Busbar

Fig. 3. 3. 5 shows E-Brake Power Supply Units(E-BPSU). There are four channel 120 V outputs, which are converted from 28 V DC through a boost DC/DC converter. The Power of each channel is 2. 5 kW.

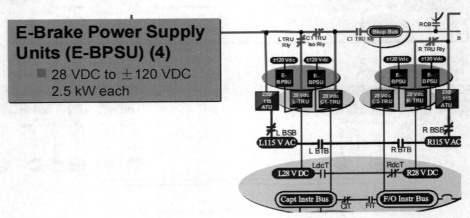

Fig. 3. 3. 5　Schematic diagram of E-BPSU

3. 28 V Low Voltage DC Busbar

According to Fig. 3. 3. 3, there are nearly 150 loads on the 28 V DC busbar, which are mainly used for fuel pump, ignite, Common Core System (CCS), Remote Distribution Controller(RDC), Busbar Power Control Unit(BPCU) or Generator Control Unit(GCU), etc. The loads with current less than 10 A are sent through the Remote Power Distribution Unit (RPDU), and the remaining load is directly sent from the 28 V DC busbar.

As shown in Fig. 3. 3. 5, there are four 28 V DC power load busbars, two of which are ordinary busbars namely L 28 V DC and R 28 V DC, the other two are emergency busbars, one of which is the captain instrument busbar (Capt Instr Bus), and the other is the flight/control instrument busbar (F/O Instr Bus) for the pilot instrument and flight instrument.

4. Emergency Power Bus

As shown in Fig. 3. 3. 5, the B787 emergency PSS has an emergency DC battery and an AC generator driven by Ram Air Turbine (RAT), called RAT AC generator, which generates 230 V variable frequency AC with a power of 10 kV · A as electrical/hydraulic emergency power supply. There are two main batteries and APU batteries, all of which are 48 A · h lithium-ion batteries. These two sets of batteries are connected to the Hot Battery Busbar(HBB). The batteries are charged by the captain's instrument emergency busbar through the Battery Charger (BC). The two chargers are made of the principle of DC/DC boost converter. The negative battery is connected to the Current Return Network (CRN).

When the main power fails, the battery supplies power to the pilot's instrument emergency busbar. Another APU battery supplies power to the flight control instrument busbar.

Arrange the power supply data of the busbar of B787 as listed in Tab 3. 3. 2.

Tab 3. 3. 2 Busbar of B787

SN	Name	Number	Main technique data
1	230 V AC bus	4	230 V AC bus L1 230 V AC bus L2 230 V AC bus R1 230 V AC bus R2
2	230 V AC connection bus	2	LTB RTB
3	230 V AC emergency bus	1	Bkup bus
4	115 V AC bus	2	115 V AC bus L 115 V AC bus R
5	270 V AC bus	4	±270 V DC bus L1 ±270 V DC bus L2 ±270 V DC bus R1 ±270 V DC bus R2
6	28 V AC bus	2	28 V DC bus L 28 V DC bus R
7	Captain instrument bus	1	Capt Instr bus 28 V DC
8	Flight/control instrument bus	1	F/O Instr bus 28 V DC
9	Main storage battery heat bus	2	Main BAT Hot bus APU BAT Hot bus
10	Wheel brake bus	4	±130 V DC bus

As shown in Fig. 3. 3. 4 and Fig. 3. 3. 5, the main 230 V AC busbar of B787 is independently powered and cannot be connected to each other in parallel. When the four variable frequency alternators are powered normally, all BTB connected to the 230 V AC busbar must be disconnected. For example, once AC starter/generator L1 or AC starter/gengrator L2 fails, the L2 BTB can be connected and the normal generator supplies power to the power busbar of the failed generator.

When the aircraft stops on the ground and starts the APU, the Aux Gen L supplies power to 230 V AC bus L2 and 230 V AC bus R1 through LAPB, L2 BTB, and R1 BTB. And the Aux Gen R supplies power to 230 V AC bus L1 and 230 V AC bus R2 through RAPB, L1 BTB, and R2 BTB. If the main generator fails, the APU generator can supply power to the 230 V AC busbar of the faulty generator. When the APU generator is power, the corresponding GCB must be disconnected first.

If both the main generator and the APU generator fail, and the RAT GEN is need to provide power supply, C1 TRU ISO Rly and BB ISO Rly of the AC emergency busbar Bkup should also be disconnected and the RAT generator only feeds Bkup bus, C1 - TRU and C2 - TRU power supply. these electrical equipment is critical for safe flight and landing.

B787 onboard ground power sockets for four electrical equipment, two of which are connected with the 115 V AC Bus, can also provide power for 230 V AC load through ATU, left and right ground power sockets are Left External Plug (LEP) and Right External Plug (REP), and external power sockets for starting APU and main generator are PESP and SESP.

The BPCU is mainly used for monitoring main power supply, APU generator, RAT generator and ground power supply. BPCU communicates with GCU of 4 generators and 2 APU generators through the data bus. It is also used to monitor the Phase Sequence(PS) and power quality of the ground power supply. Only when it is consistent with the requirements of the aircraft network, the ground power contactors are connected.

3.4 Aviation Battery

3.4.1 Application Overview of Aviation Battery

Under normal circumstances, the generator supplies power to the electrical equipment and charges the battery at the same time. When the generator fails, power is supplied by batteries. The main use of aircraft battery is when the aircraft main power supply does not work or failure as an auxiliary power supply or emergency power supply to important electrical equipment power supply, or as a power source to start aircraft engines. In some special cases (such as in the field without power supply), it can also be used as a power source for pre-flight inspection of some small power electrical equipment. Small and medium-sized aircraft batteries used as emergency power supply, often used in parallel with the DC generator (main power supply).

The types of aviation battery include lead acid battery, nickel-cadmium battery, silver-zinc battery and lithium-ion battery. Among them, nickel-cadmium battery and lithium-ion battery are being used in modern aircraft, while others have gradually faded out of the market. Another new clean cell, fuel cell, is still in its infancy in aviation and not in many applications.

Nickel-cadmium battery is used in aviation due to their long life and convenient maintenance. For example, Airbus 320 series aircraft. Compared with lithium-ion battery, nickel-cadmium battery is bulky, heavy, lack of charge and discharge current, and slow to charge.

The plug cover of nickel-cadmium battery is relatively special, which can not only discharge the gas formed in normal use, but also make the electrolyte not overflow during the maneuvering flight. In order to ensure the normal operation in the low temperature environment, the nickel-cadmium battery is equipped with a heating or thermal insulation device.

In addition, aircraft are often equipped with self-powered systems such as emergency locator beacons, life rafts and life jackets. Lithium is one of the alkaline active metals and one of the lightest elements, which makes its use on aircraft very advantageous. Lithium-ion battery technology is a fast-growing and promising battery technology that is being used discreetly in aircraft such as smoke detectors, engine start-up and emergency backup power because of their high energy to weight ratio, no memory effect and low discharge/charge rate when not in use. Compared with lead batteries and nickel-cadmium batteries, lithium-ion batteries have many advantages, such as long life, light weight, low maintenance and short charge time. The disadvantages are high cost, flammable electrolyte, and loss of about 10% of storage capacity each year whether used or not. The aging rate of batteries is affected by temperature. The higher the temperature is, the faster the battery ages.

3.4.2 Lithium-ion Battery

Lithium-ion battery: A rechargeable lithium-ion polymer battery using colloidal or solid polymer instead of liquid organic solvent has better safety.

Rechargeable lithium-ion battery consists of graphite electrodes as a negative electrode, cobalt, manganese or iron phosphate as a positive electrode, and an electrolyte to transport the lithium ions.

Lithium-ion battery works mainly by moving lithium ions between the positive and negative electrodes. During charging and discharging, lithium ions are embedded and de-embedded between the two electrodes. When charging, lithium ions are removed from the positive electrode and embedded into the negative electrode through electrolyte, which is in a lithium-rich state. lithium-ion battery uses lithium element material as electrode, is the representative of modern high-performance battery.

1. Structure and Composition of Lithium-ion Battery

Lithium-ion battery have cylindrical and prismatic structures, as shown in Fig. 3. 4. 1. Fig. 3. 4. 1(a) shows a cylindrical lithium-ion battery, which has the advantages of easy manufacturing, excellent mechanical stability and can withstand high internal pressure. Fig. 3. 4. 1 (b) shows a prismatic lithium-ion battery, which has the advantages of small space and good heat dissipation.

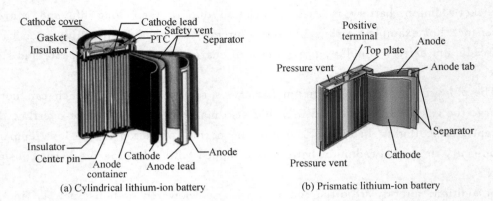

Fig. 3. 4. 1　Columnar lithium-ion battery structure

The LCO battery in Tab. 3. 4. 1 has been used for some time and is relatively mature; NCA battery have the advantages of long life and high density, NCM and LMS cells decay by the dissolution of manganese, and LMS has a more stable spinel structure.

Tab. 3. 4. 1　Chemical composition of several lithium-ion batteries

SN	Materials	Capacity/ $(mA \cdot h \cdot g^{-1})$	Nominal Voltage/V	Wh/kg	Wh/L
1	$LiCoO_2$ (LCO)	145	4. 0	602	3. 073
2	$Li(Ni0. 85Co0. 1Al0. 05)O_2$ (NCA)	160	3. 8	742	3. 784
3	$Li(Ni1/3Co1/3Mn1/3)O_2$ (NCM)	120	3. 85	588	2. 912
4	$LiMn_2O_4$ (LMS)	100	4. 05	480	2. 065
5	$LiFePO_4$ (LPF)	150	3. 34	549	1. 976

LPF battery is the latest battery, with good overcharge safety, low density, due to the low voltage of single battery, so more single battery is needed.

2. Discharge Characteristics of Lithium-ion Battery

Lithium-ion battery has a nominal voltage of 3. 6 V, and charging requires a constant voltage of 4. 2 V and associated current-limiting measures. When the battery voltage reaches 4. 2 V and the current drops to about 7% of the initial charging current, the battery is fully charged. Fig. 3. 4. 2 shows a typical discharge curve of lithium-ion battery. The effective capacity of lithium-ion battery will increase at a low discharge rate and decrease at a high discharge rate.

For security, the battery needs to be monitored and alerted based on software during charging. Lithium-ion batteries have been used in the B787, and the following factors should be considered in design and maintenance: keep the battery temperature and pressure safe, reduce the risk of explosion, prevent electrolyte overflow, disconnect the charging power when the temperature is too high, and provide an alarm for low power.

Lithium-ion batteries are not going to be replaced by aircraft main batteries all at once.

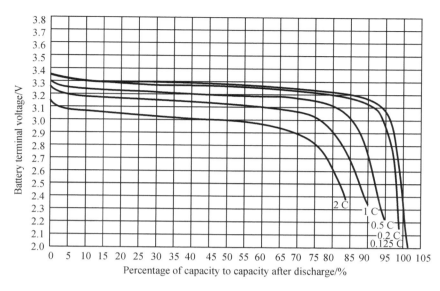

Fig. 3. 4. 2 **Discharge characteristics of lithium-ion batteries**

Both aircraft and batteries need safety measures. These safety measures include circuits and hardware to keep voltage and current within safe limits.

3. Main Features of Lithium-ion Batteries

① High energy density. The energy density varies with the electrode materials, up to 150—200 W · h/kg (540—720 kJ/kg) by mass, or 250—530 W · h/L (0. 9—1. 9 kJ/cm³) by volume.

② High open circuit voltage. It varies with different electrode materials, up to 3. 3— 4. 2 V.

③ Large output power. Different electrode materials, up to 300—1 500 W/kg(@20 s).

④ No memory effect. Lithium-ion phosphate battery has no memory effect, the battery can be charged and discharged at any time without empty power, easy to use and maintenance.

⑤ Low self-discharge. The self-discharge rate was less than 5%/month to 10%/month. The intelligent lithium-ion battery system with the internal detection circuit can monitor the charge and discharge condition to avoid the occurrence of overcharge and over discharge.

⑥ Wide operating temperature range. Lithium-ion batteries can work properly in between—20—60 ℃.

⑦ Fast charging and discharging speed.

4. Disadvantages of Lithium-ion Battery

Unlike other rechargeable battery, the capacity of lithium-ion battery declines slowly, not with the number of times they are used, but with the temperature. The likely mechanism is a gradual increase in internal resistance, so it is easier to manifest in electronic products with high operating current.

5. Lithium-ion Battery on the B787

Boeing B787 adopts advanced lithium-ion battery. Fig. 3. 4. 3 shows the installation of B787 lithium-ion battery on the plane. Due to the use of lithium-ion battery, weight can be saved by 30%. In fact, B787 has two groups of lithium-ion batteries, one before and one after, each weighing about 28. 5 kg, each containing eight 4 V-cells, after series connect reached 32 V output voltage, can provide 75 A · h of electrical energy.

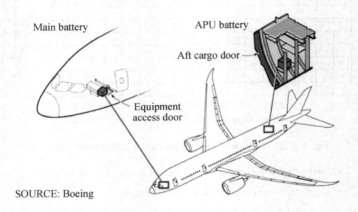

Fig. 3. 4. 3   Lithium-ion battery installation in B787

Lithium-ion battery incident: On January 7, 2013, an auxiliary power battery in the rear of a JAL B787 overheated on the auxiliary power battery at Boston Logan Airport, causing it to catch fire.

It took the airport fire brigade more than an hour to extinguish the flames. A subsequent inspection revealed that not only were the batteries and shells badly damaged, but the leaking molten electrolytes and hot gases had also damaged the structure of the body half a meter away. The National Bureau of Traffic Safety Investigation (NBTSI) survey found that local steel structures showed signs of condensation after gasification, indicating that local temperatures could be as high as up to 3 000 degrees. Nine days later, on January 16, another full-Japanese B787 flew from Yamaguchi to Tokyo Narita Airport. At cruise altitude shortly after takeoff, the pilot smelled pungent smoke in the cockpit and warning lights on the dashboard also showed a battery failure. The plane immediately made an emergency landing at Takamatsu airport, but the 129 passengers and eight crew members on board escaped safely through the emergency exit and the inflatable slide. The later inspection found that the main battery in the electronic module under the front fuselage cockpit had overheated and burned, and the shell was seriously damaged.

3.4.3 Whether Fuel Cell Can Be Used as Future Aviation Battery

Fuel cell is a power generation device that converts the chemical energy of fuel directly into electricity.

The principle of fuel cell is an electro-chemical device, and its composition is the same as that of ordinary batteries. The cell consists of two electrodes (fuel and oxidant) and an electrolyte. The difference is that the active substances in ordinary batteries are stored inside the battery, thus limiting the capacity of the battery. The positive and negative electrodes of fuel cell do not contain active substances themselves, only catalytic conversion elements. Fuel cell is therefore veritable energy-conversion machines that convert chemical energy into electricity. When the battery works, fuel and oxidizer are supplied externally to react. In principle, a fuel cell can generate electricity continuously as long as the reactants are being fed in and the products are being removed. The following uses hydrogen-oxygen fuel cell as an example to illustrate the operating principle of fuel cell.

1. Introduction of Hydrogen-Oxygen Fuel Cell

Fig. 3.4.4 shows the schematic structure of hydrogen-oxygen fuel cell.

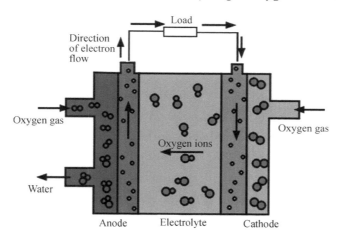

Fig. 3.4.4 Schematic diagram of hydrogen-oxygen fuel cell

The anode feeds hydrogen, the cathode feeds oxygen, and the electrolyte has hydroxide ions. Electrolysis of water involves the loss of four electrons for every two water molecules converted into two hydrogen molecules and one oxygen molecule. Their equation is as follows:

Cathode: $2H^+ + 2e \rightarrow H_2$

Anode: $4OH^- \rightarrow O_2 + 2H_2O + 4e$

Equation: $2H_2O \rightarrow 2H_2 + O_2$

The reaction principle of hydrogen-oxygen fuel cell is the reverse process of electrolysis of water. Every 2 water molecules can provide 4 electrons to the external circuit at the same time, and their reaction formula is as follows:

Cathode: $H_2 + 2OH^- \rightarrow 2H_2O + 2e$

Anode: $O_2 + 2H_2O + 4e \rightarrow 4OH^-$

Equation: $2H_2 + O_2 \rightarrow 2H_2O$

In addition, only the fuel cell itself can not work. It must have a corresponding auxiliary system, including the reaction agent supply system, heat removal system, drainage system, electrical performance control system and safety device, etc.

2. Types of Fuel Cell

There are currently six categories of fuel cells.

① Polymer Electrolyte Fuel Cell (PEFC), also known as Proton Exchange Membrane (PEM) battery, operating temperature is 60—120 ℃, used for transport aircraft demonstration and small power occasions, power level is 5—250 kW.

② Alkaline Fuel Cell (AFC), whose operating temperature is less than 100 ℃, have been used in the aerospace field since 1960.

③ Phosphoric Acid Fuel Cell (PAFC), with operating temperature 160—220 ℃, are widely used in airports, hospitals and schools.

④ Molten Carbonate Fuel Cell (MCFC), with operating temperature 600—800 ℃, are only applicable to large-scale power plants.

⑤ Direct Methanol Fuel Cell (DMFC), operating temperature is 60—120 ℃, suitable for portable equipment power supply.

⑥ Solid Oxide Fuel Cell (SOFC), operating temperature is 800—1 200 ℃, in heavy locomotive APU, aircraft and other high power devices have been applied.

Especially like Proton Exchange Membrane Fuel Cell (PEMFC), because it has many advantages over battery and other power generation devices, as a equipped power source has received a lot of attention.

The advantage of fuel cell is that they can generate electrical power as long as fuel is supplied. They are zero-emission, quiet, and have no moving or rotating parts (high reliability).

The disadvantages of fuel cell are long start-up time and slow dynamic response speed. The dynamic response time of the battery is one thousandth of a second, and it can reach 1 000 charge and discharge cycles. However, compared with the fuel cell, it has the disadvantages of low capacity and short discharge time.

3. Fuel Cell System Description

According to the previous studies, the main advantage of fuel cell is high energy density, while the main disadvantage is slow dynamic response speed. The battery has high power grade and output, and a fast dynamic response. Therefore, these two kinds of batteries are often used together to form a hybrid PSS, which makes it more attractive.

The block diagram of the hybrid PSS is shown in Fig. 3. 4. 5. It is a system composed of fuel cell via DC/DC converter and battery.

Fuel cell is electro-chemical devices that convert chemical energy directly into electricity. Among the various types of fuel cells, PEF/PEM cells are the most attractive

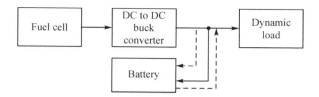

Fig. 3. 4. 5   Block diagram of hybrid power supply

due to their robustness, simple structure, fast start-up, high power density and low temperature operation characteristics.

If fuel cell can be used in aircraft, with some design compromises, proton exchange membrane fuel cell could be used to power aircraft aisles and in-flight entertainment systems.

The main advantage of PEM fuel cell having a higher power density than any other type of fuel cell, except AFC, is their high power density.

The use of solid electrolytes has many advantages over liquid electrolytes, such as no monitoring of electrolyte corrosion and status. Solid-state electrolytes, which allow fuel cell to operate at a lower temperature, are useful in situations where quick start-up is required. In addition to high energy density and fast start-up, PEM fuel cell has the advantage of long life.

Fuel cell has a wide range of applications, including military, space, power plants, motor vehicles, mobile devices, and households. Early fuel cell development focused on specialized applications such as military space and distributed power generation above kilowatts. The field of electric vehicle has become the main direction of fuel cell application, and there are a variety of automatic vehicles using fuel cell power generation in the market. In addition, through the miniaturization technology will be used in fuel cell in general consumer electronics is one of the application development direction, in the progress of technology, the future miniaturization of fuel cell will be available to replace the existing lithium-ion battery or nickel metal hydride battery high-value products, such as used in notebook computers, wireless phone, VCR, camera and so on to carry the power of the electronic products. In the past 20 years, fuel cell has experienced several development stages, such as alkaline, phosphoric acid, molten carbonate and solid oxide, and the research and application of fuel cell is developing at a very fast speed. Alkaline Fuel Cell (AFC) are the fastest growing of all fuel cell and are used to power and drink water for space missions, including the Space Shuttle. Proton Exchange Membrane Fuel Cell (PEMFC) has been widely used as transportation power and small power supply device. Phosphoric Acid Fuel Cell (PAFC) has entered the commercial stage as a medium power source and is the first choice of civil fuel cell. Molten Carbonate Fuel Cell (MCFC) have also completed industrial trials. Solid-Oxide Fuel Cell (SOFC), which started late, is the most promising fuel cell in the field of power generation and is the preferred object of large-scale clean power stations in the future.

For many years, people have been trying to find a way of energy utilization that has high energy utilization efficiency and does not pollute the environment, and fuel cell is a relatively ideal power generation technology.

Fuel cell is very complex, involving chemical thermodynamics, electro-chemistry, electro-catalysis, materials science, power system and automatic control and many other disciplines related to the theory, with high power generation efficiency, less environmental pollution and other advantages.

3.5　Summary

This chapter introduces the radical change of the aircraft PSS from the generator, the main power supply, the auxiliary power supply and the emergency power supply and other aspects are introduced.

① The PSS develops from low-voltage DC power supply to the CF Aircraft PSS system, and then to variable frequency AC PSS. In the future, advanced aircraft will develop towards high-voltage AC PSS and high-voltage DC PSS.

② B787, the representative of advanced aircraft, is closer to AEA. Its generator is a variable frequency generator, and its operating frequency changes from 360 Hz to 800 Hz, greatly reducing its volume and weight.

③ Due to the adoption of more-electric technology, its PSS has undergone fundamental changes, and there are many problems worth studying in the main power supply, auxiliary power supply and secondary power supply.

④ More-electric/all-electric batteries are introduced. Lithium-ion batteries have been applied in MEA B787, but there are fire hazards in lithium-ion batteries. Nickel-metal hydride batteries and fuel cell are attracting attention. The power system of the future may be a hybrid power system using a combination of fuel cell and batteries.

3.6　Exercises

1. What are the development stages of the AC power system? What are the characteristics of the AC power supply at each stage?

2. What is a narrow frequency conversion AC power supply system? What is a wide frequency AC power system?

3. What are the shortcomings of the frequency conversion AC power supply?

4. What does the load not adapt to the frequency conversion AC power supply?

5. What is the role of the aviation battery on the plane?

6. What are the advantages of a lithium-ion battery compared with traditional batteries? What are the shortcomings?

Chapter 4 Power Distribution System of MEA

With the development of the aviation industry to MEA/AEA, the aircraft demand for electrical energy has increased, and the use of aircraft electrical energy management means to reduce the loss of power grid energy and optimize the overall efficiency, etc. , is more conducive to the overall performance of the aircraft. The power distribution system is an important link between the power supply and the power-using equipment, and is the basis for achieving fault tolerance and uninterrupted power supply. However, it is impossible to completely eliminate the fault of power supply, grid conductor and power-using equipment, and it is crucial that the power supply to power-using equipment should still be reliable under the fault condition.

In the case of a twin-engine aircraft, for example, the requirements for the electrical distribution system are that for critical electrical equipment, the PSS must have the ability to still operate in the event of two failures. For the main electrical equipment, the PSS must have the ability to operate even after one failure, unless the Starter/Generator(S/G) of that engine fails, the PSS still has the ability to start the engine when one failure occurs, one failure in the electrical system or one generator failure should not affect the normal operation of the aircraft's electrical equipment, and the failure of the power supply and distribution system should not expand on its own and should not cause unsafe factors.

The structure of the MEA power distribution system improves the maintainability, reliability, flight safety, and efficiency of the aircraft. This section describes the electrical equipment, distribution methods, and electrical load management of MEA.

4.1 Electricity Consumption Equipment

Electrical equipment is a device that uses electricity to work. Due to the different functions of electrical equipment in the flight process, the impact on flight safety is different. According to the different importance of electrical equipment, the electrical equipment is divided into different categories to realize the classification of power supply priority.

4.1.1 Load Classification

Aircraft electrical equipment is divided into three categories: general electrical loads, major electrical loads and important electrical loads. Important electrical loads are the minimum electrical loads necessary for safe flight and landing, major electrical loads are the loads required to complete the flight mission, and general electrical loads are the loads that have less to do with flight safety and completion of the flight mission.

4.1.2 Characteristics of MEA Power Equipment

With the continuous equipment of power electronics technology in aircraft, the large number of power electronic converters and electric motors has led to important changes in the electrical equipment of aircraft, mainly in:

① Electronic device power supplies and motor controllers are power electronic conversion devices with non-linear, constant power, pulse time-varying, energy regeneration, capacitive input impedance and other characteristics, which bring new challenges to the quality and stability of the PSS.

② Electronic devices with internal power supply and motors with speed control are typical of constant power devices, where the supply voltage increases and the consumption current decreases.

③ Mono-phase static converter is a typical time-varying load for power supply. If the converter feeds the resistive load, the load power varies twice as much as the converter output power, with the maximum value being double the rated power of the load and the minimum value being almost zero, and the power varies even more sharply if the converter feeds rectified load.

④ When the motor braking operates in the aircraft mechanical and electrical driving mechanism operates, the mechanical energy of the movement part will be fed to the power supply, resulting in the increase of the aircraft power grid voltage. The aircraft power grid should have the ability to quickly absorb the regenerative energy.

⑤ Both converter and motor controllers have capacitive input filters. The inrush current when capacitive loads are turned on may cause the SSPC to malfunction, and a large number of capacitive loads turned on at the same time will cause a sharp drop in grid voltage.

In recent years, many technological advances have been made in the generation, switching and protection of an electrical power that are beginning to have an impact on classical electrical systems, with new devices, materials and theories emerging.

4.1.3 MEA Electric Load

The main civil MEA are A380 and B787, which are briefly introduced below.

1. A380 Electric Load

(1) Cockpit Electrical Load

Tab. 4.1.1 lists the electrical loads in the cockpit, most of which are critical loads that require emergency power to complete the mission and are related to the safety of the aircraft.

Tab. 4. 1. 1　Cockpit electrical load

SN	Load type	Electrical load parameters				
		Voltage/V	Frequency/Hz	Characteristic	Phase NUM.	Load importance1
1	Radio navigation equipment(call)	115	400	AC	1	Importance electrical loads
2	Radar equipment	115	400	AC	3	Importance electrical loads
3	Gyroscope meter	36	400	AC	3	Importance electrical loads
4	Electric instrument, rotation instrument, autopilot	36	400	AC	3	Importance electrical loads
5	Instruction system	115	400	AC	1	Importance electrical loads
6	De-icing facilities	115	Electric pulse	AC	1	Importance electrical loads
7	Flight Control Computer(FCC)	115	400	AC	1	Importance electrical loads
		Start-up or power supply method				Nature of load
8	EHA	Aviation battery start				Importance electrical loads
9	Drive pumps	Aviation battery start				Importance electrical loads
10	Auxiliary power unit	Aviation battery start				Importance electrical loads
11	Flight data recorder	Aviation battery				Importance electrical loads

As shown in Fig. 4. 1. 1, the cabin of the A380 aircraft is schematically divided into four parts, namely the cockpit, the passenger cabin, the upper and lower cargo, and passenger convenience parts, for the convenience of analysis.

1—Cockpit; 2—First class; 3—Lounge bar area; 4—Business class; 5—Economy class; 6—Shop

Fig. 4. 1. 1　A380 cabin diagram

(2) Electrical Loads for Passenger Compartment and Upper and Lower Cargo Compartment

The electrical loads of passenger compartment and upper and lower cargo compartments are listed in Tab. 4. 1. 2. The total volume of the upper and lower cargo compartments is 1 570 m³, and the total area of the two passenger cabins is 550 m². The A380 has 15 different temperature control zones with a volume of 1 500 m³, and the air is changed every 3 minutes with a temperature adjustment range of 18—30 ℃. The power consumption of multimedia entertainment facilities, 70-seat first class and 428-seat economy class and 8 luxury first class compartments, where the luxury compartments refer to the power consumption parameters of four-star hotel rooms. The cabin lighting adopts fluorescent and LED lighting technology, with 5 000 lighting scenarios to choose from, and five applicable lighting scenarios for the A380 type of aircraft currently in service in China.

The oxygen supply system can refer to the air-conditioning electricity consumption, but it must be separated from the air-conditioning system, serving as a backup. Kitchen power, there are 19 kitchens on the A380 to provide passengers with in-flight food, with reference to kitchen power, only one heating microwave oven, one fresh refrigerator, one coffee machine and one oven are considered.

Tab. 4. 1. 2　Electrical loads in passenger compartments and upper and lower cargo compartments

SN	Load type		Electrical load parameters				
			Voltage/V	Frequency/Hz	Characteristic	Phase NUM.	Importance
1	Central air-conditioning		115	400	AC	1	Main electrical loads
2	Multimedia interface		115	400	AC	1	Secondary electricity load
3	Lighting		36	400	AC	3	Main electrical loads
4	Oxygen supply system		36	400	AC	3	Important electrical loads
5	Kitchen electricity	Heating microwave oven	115	400	AC	1	Main electrical loads
		Refrigerator	115	400	AC	1	Secondary electricity load
		Coffee machine	115	400	AC	1	Secondary electricity load
		Ovens	115	400	AC	1	Secondary electricity load

(3) Electricity Load for Passenger Convenience Facilities

As listed in Tab. 4. 1. 3, it is the load table for passenger amenities, where bar power refers to bar equipment power; toilet electricity refers to the water supply system, lighting system, the toilet in the pumping motor electricity and ventilation system; warning signal and exterior lighting system refer to the load that contains all sensor power, cabin door power, warning system, cabin power marker light and exterior lighting.

Tab. 4. 1. 3　Electricity load for passenger convenience facilities

SN	Load type	Load Name	Voltage/V	Frequency/Hz	Characteristic	Phase NUM.	Importance
1	Electricity for bars	Refrigerator system	115	400	AC	1	Secondary electricity load
		Lighting system	36	400	AC	3	Main electrical loads
		Sound System	36	400	AC	3	Secondary electricity load
2	Electricity for box stores		—	—	—	—	Secondary electricity load
3	Electricity for elevators		115	400	AC	1	Main electrical loads
4	Electricity for warning signals		28	—	DC	—	Main electrical loads
5	Exterior lighting system		115	400	AC	1	Main electrical loads

(Continued)

SN	Load type	Load Name	Voltage/V	Frequency/Hz	Characteristic	Phase NUM.	Importance
6	Electricity for the bathroom	Water supply system	36	400	AC	3	Main electrical loads
7		Lighting System	36	400	AC	3	Main electrical loads
8		Pumping system	36	400	AC	3	Secondary electricity load
9		Ventilation system	36	400	AC	3	Main electrical loads

2. B787 Electrical Load

The electricity consumption loads are classified in Tab. 4. 1. 4.

Tab. 4. 1. 4　Electricity consumption load classification

SN	Voltage	Quantity/Unit	Typical load	Remark
1	230 V AC	40	Wing anti-icing, aft galley equipment, cargo hold heating, fuel pumps ECS fans, etc.	High power
2	115 V AC	20	Cooling fans, cockpit heating, front and middle galley equipment	Greater than 10 A
3	28 V DC	150	DC fuel pump, engine ignition, GCU and BPCU, flat panel display, common core system	Greater than 10 A
4	115 V AC, 28 V DC	900	Conventional aircraft load	Less than 10 A
5	270 V DC	11	Electric hydraulic pump (4), ECS compressor (4), electric ventilator (2), nitrogen generator (1), starter/generator (4+2)	—

4.2　Structure of MEA Power Distribution System

The distribution of the aircraft power grid is determined by the location of the electrical equipment on the aircraft, which is distributed over almost the entire body of the aircraft. Therefore, the aircraft power grid is very complex. It is more easy to occur in the wire short circuit, open circuit or other faults, and there are some specific requirements for the aircraft power grid.

4.2.1　Requirements for Aircraft Power Grids

The requirements of the aircraft power grid are mainly reflected in the following aspects:

① The aircraft power grid must have high reliability and strong vitality. This requires that under normal operation and various fault conditions, the power supply is guaranteed without interruption, especially to ensure the continuous power supply of important power equipment for safe return to flight.

② The quality of power supply should be high. The quality of power supply directly affects the performance of electricity load.

③ The weight of the grid should be light. Low-voltage DC power supply has low voltage and high current, thus the wire is thick and heavy, and it is of great significance to reduce the weight of the grid.

④ The aircraft grid should also be easy to install, inspect, repair and modify.

Early aircraft had a wooden structure and a double-line power grid. The fuselage of modern aircraft is mostly composed of metal skeleton and skin, with special metal woven wire to connect all parts of the body, forming a low resistance path, so the body can be used as a negative loop of the power grid, so that the aircraft DC electric network can be made of single wire.

The single wire system has the following advantages: only the circuit positive polarity with the wire, the power grid weight is light. It reduces the number of wire connections and switching equipment, improves the power supply quality and simplifies the power grid structure; it eliminates the electrostatic induction between the wire and the metal body.

The main disadvantages of single-line system are: easy to occur short circuit, the magnetic field formed by a single wire is strong, such as unreasonable distribution will affect the work of magnetic compass and other equipment.

4.2.2 Power Distribution Method

Fig. 4. 2. 1 shows a comparative diagram of two power distribution systems, where Fig. 4. 2. 1(a) shows a centralized power distribution system for conventional aircraft. Although the capacity of the power grid is large, it is gradually faded out due to its complex cockpit equipment, heavy weight of distribution conductors and easy spread of faults. And

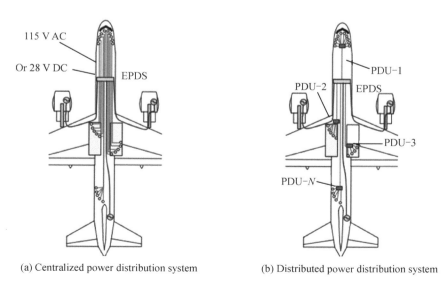

(a) Centralized power distribution system   (b) Distributed power distribution system

Fig. 4. 2. 1   Comparison chart of the two power distribution systems

Fig. 4. 2. 1(b) is the distributed power distribution system, which is commonly used in the power distribution system of modern aircraft. As the MEA is characterized by variable frequency conversion AC power supply and 270 V high voltage DC, its distribution system adopts distributed power supply system, and the key distribution device is Remote Power Distribution Unit(RPDU), which greatly reduces the weight of the distribution wire, and the fault is easy to be isolated, but the control logic is complex.

The MEA distribution system is divided into Primary Distribute Power System(PDPS) and Secondary Distribute Power System(SDPS), as shown in Fig. 4. 2. 2. The integrated display device of the power system communicates with the BPCU, the power controller of the sink, through the ARIN429 bus. The power distribution information from the PSS sends power supply request and stops power supply information to the contacts through the BPCU. Based on the information from each contact, the secondary power distribution system sends load access or disconnects information to the Electrical Load Management Center (ELMC) via the ARIN429 bus.

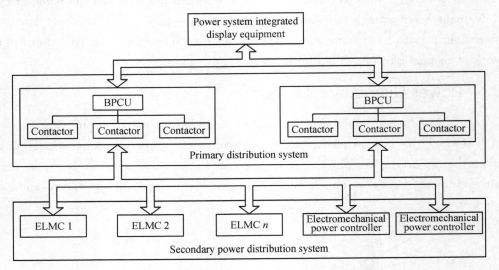

Fig. 4. 2. 2　Aircraft power distribution system topology

4.2.3　Busbar Power Controller Unit

The Busbar Power Controller Unit (BPCU) consists of an internal power supply, computer, interface circuit, conditioning circuit and communication interface for monitoring, controlling and protecting the power supply and power grid, with the following functions:

① Ground power monitoring.

② APU or Integrated Power Unit (IPU) start-up.

③ Electric starting of the main engine.

④ Input, removal and load equalization of the main generator (in synergy with GCU).

⑤ Fault monitoring, diagnosis and protection of parallel power supplies.

⑥ Statistics of the total electrical load and coordination with power supply operation.

⑦ System self-test, operation status record, fault record, sending monitor to show operation status, fault information and ground maintenance information.

⑧ Communication with GCU and remote distribution center, communication with higher-level computers, etc.

4.2.4 Electrical Load Management Center (ELMC)

Developed and manufactured by General Electric Aviation, the B777 aircraft electrical load management system breaks new ground in electrical load management with the layout shown in Fig. 4.2.3, the first integrated electrical power distribution and load management system for civil aircraft.

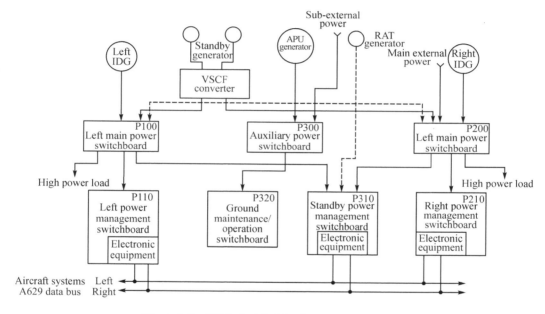

Fig. 4.2.3　B777 electrical load management system

1. Power Distribution Board

Distribution boards are divided into primary power distribution board and secondary power distribution board. The details are as follows:

(1) Primary Power Distribution Board

The system includes seven power distribution panels, three of which are related to primary power distribution, namely.

① P100: The left main power distribution panel distributes and provides protection for the left main power load.

② P200: The right main power distribution panel distributes and provides protection for the right main power load.

③ P300: Auxiliary power distribution panel distributes and provides protection for the main power load of auxiliary equipment.

(2) Secondary Power Distribution Board

Secondary power distribution is undertaken by four secondary power distribution panels, which are.

① P110: Left power management switchboard for load distribution, power supply protection and control related to the left channel.

② P210: Right power management switchboard for load distribution, power supply protection and control related to the right channel.

③ P310: Standby power management switchboard for load distribution, power supply protection and control related to the standby channel.

④ P320: Ground maintenance/operation switchboard distributes and protects the power supply associated with ground operations.

Load management and general system control are performed through electronic Equipment Unit (EU) installed in the P110, P210, and P310 power management switchboards. Each of the equipment unit interfaces with the left and right aircraft systems ARINC629 digital data bus and is constructed with two remainders for attendance reasons. The EUs include a set of modular Line Replaceable Unit (LRU). When the chassis cover is opened, the circuit board modules can be quickly replaced. The highly modular structure reduces the risk of development and results in a mean time between failures of 200 000 hours.

The ELMC is also known as the RPDU. As shown in Fig. 4.2.4, the B777 electrical load management system. ELMC consists of distribution busbars, secondary power supply, SSPC or contact, and Electronic and Electrical Controller (EEC), etc. The EEC makes ELMC a smart distribution box.

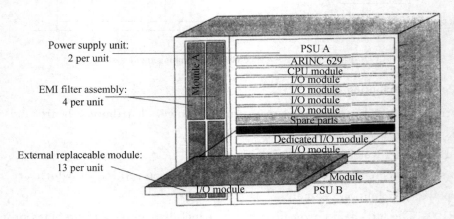

Fig. 4. 2. 4   The B777 electrical load management system

ELMC internal busbar has 270 V DC, 28 V DC, 115 V/400 Hz mono-phase busbar and three-phase busbar, and the operating power of distribution box is 270 V.

2. Main Functions of the Distribution Board

The distribution board assumes the role of safely conveying electrical energy to the entire electrical system, specifically as follows:

① Distribution to the electrical equipment. The ELMC delivers power to the load through the SSPC with the help of the EEU and enables fault detection and protection of the distribution line.

② Electrical energy converter to convert 270 V DC to 28 V DC or 115 V 400 Hz AC.

③ Automatic switching of AC power failure.

④ Energy management of the battery to keep the battery in a fully charged state and the battery to supply power to critical power-using equipment in emergency situations.

⑤ Priority management of electronic devices. The electronic equipment is divided into 3 levels according to its importance. When all the equipment of the PSS is normal, power can be supplied to various electronic equipment. When one or two generators fail, the auxiliary power should be connected immediately, and at the same time, some loads should be removed according to the importance of the electronic equipment. The priority management of electronic equipment is to try to get power to the equipment that must be supplied under the condition that the generator is not overloaded.

⑥ Communication function. The ELMC is capable of receiving commands from the higher-level computer and flight crew, and reporting its own operating status and operating parameters, and reporting fault information to the higher-level computer.

⑦ Fault Prediction and Warning. The MEA is equipped with numerous electronic and electrical devices, and these electronic components are prone to failure in adverse environments, such as high and low temperatures and temperature cycling, low air pressure, humidity, and incorrect design. The failure of a single point or a unit will cause the failure to spread if the faulty part is not removed in time, and it is necessary to eliminate the failure before it occurs through failure prediction and early warning. The quality of electrolytic capacitors in power electronics devices, for example, is not only affected by the temperature of the operating environment, but also with a significant reduction in capacity at low temperatures. In addition, as the operating time increases, the capacity decreases and the equivalent series resistance ESR increases, which leads to an increase in the voltage pulsation on them, so the operating condition of the capacitors can be known by testing their parameters.

⑧ System self-test and fault diagnosis, protection and alarming. For example, the faults in ELMC are faults in EEU, faults in power supply of EEU, faults in secondary power supply, faults in SSPC, faults in contact electrical device and faults in busbar. Even with a fault prediction system, faults in the ELMC cannot be avoided. Therefore, it is necessary to detect faults in time, implement protection and prevent faults from spreading and expanding.

⑨ Electromagnetic Compatibility(EMC). The ELMC is equipped with various modules inside, as shown in Fig. 4. 2. 4, which shows various I/O modules, CPU modules, ARINC429 communication modules and power supply modules. These modules are co-located in the chassis and should have good EMC characteristics.

3. Advantages of Using ELMC

An electrical load management center has the following advantages:

① The ELMC built-in electrical and electronic controller EEU to realize the detection and switch with the internal busbar.

② Automatic management of loads.

③ Distribution network fault detection and protection to improve the performance and maintenance of aircraft PSS.

④ Reporting of trouble shooting and power consumption to the PSS processor.

⑤ Monitoring the quality of the power supply of the incoming power and protecting equipment with high power consumption quality.

⑥ The use of redundant power input to achieve reliability of power supply to critical on-board power equipment.

⑦ System reconfiguration of secondary power distribution.

⑧ The use of an electrical load management to centrally manage the aircraft's local electrical equipment centrally can reduce the length of power lines inside the airframe, thereby reducing the weight of the aircraft's electrical system.

Therefore, the ELMC is an intelligent power distribution device of the automatic power distribution system, whose role is equivalent to that of a conventional aircraft switchboard. The built-in EEC of ELMC realizes the detection and switching of internal busbars, priority management of loads, distribution network fault detection and protection, and reports the fault handling and power consumption to the PSS processor, accepts control instructions from the higher-level computer and responding to it. In order to improve the quality of power supply, reduce the weight of the grid and monitor the whole electrical network, the secondary power supply should be designed in the vicinity of the ELMC, or the secondary power supply can be structurally combined with the ELMC to manage the redundant secondary power supply with the help of the EEU of the ELMC.

The load management system control characteristics provided by ELMS are far ahead of any equivalent system in use on the airline today. There are approximately 17—19 electrical load control units depending on the aircraft configuration that supply power and control loads directly from the aircraft AC Main Busbar(MBB). These loads are controlled by intelligent functions included in the EU of the ELMS. The main advancement is the advanced load disengagement and load optimization function, which tightly controls the functional availability of each unit if the main power supply fails or cannot be applied. The system can reconfigure the load to achieve the best distribution of available power. In case of power restoration, the system can restore the load in many arrangements. So the system can optimize the utilization of power at all times instead of taking off the load in case of an emergency.

The use of ELMS has led to a significant reduction in size, wiring and plugs, weight, relays, and circuit breakers.

The built-in intelligence, the application of a digital data bus, the maintenance features and the extensive in-flight testing have reduced the system's manufacturing and in-flight testing time to about 30% of that of contemporaneous systems.

4.2.5 Secondary Power Distribution

Secondary power distribution plays the role of direct power supply to each load. In order to compare the secondary power distribution of traditional aircraft and modern advanced aircraft, it is introduced as follows respectively:

1. Secondary Power Distribution with Cockpit Circuit Breaker Panel

Conventional aircraft often use the cockpit secondary power distribution method of circuit breakers, as shown in Fig. 4. 2. 5. The white number on each circuit breaker indicates the rated value of current allowed to pass, for example, "5" is rated at 5 A, when the current flowing through the circuit breaker reaches or exceeds 5 A, the circuit breaker trips and cuts off the circuit.

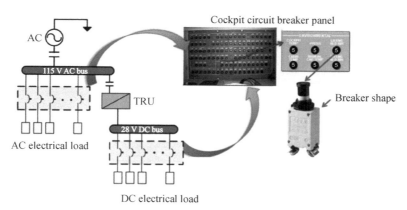

Fig. 4. 2. 5 Secondary power distribution with cockpit circuit breaker panel

As can be seen from Fig. 4. 2. 5, if the number of secondary electrical loads is large, a very large panel in the cockpit is required to install circuit breakers and cause the cockpit to be very crowded.

The trigger characteristic curve of the circuit breaker will be very important to the safety of the circuit and even the safety of the whole power grid. Usually, a trigger characteristic with inverse delay is required, that is, the larger the Over Current(OC), the shorter the trigger delay, and the smaller the over current, the longer the trigger delay. Fig. 4. 2. 6 shows the over current triggering characteristics of a circuit breaker at 25 ℃.

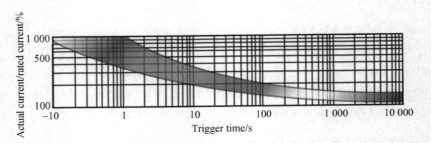

Fig. 4. 2. 6　Trigger characteristic curve (trigger time, current characteristic (25 ℃))

2. Secondary Power Distribution Using Integrated Power Distribution Units

Usually, the main power supply of the aircraft is 115 V AC and 28 V DC generated by 115 V through the converter TRU unit. Therefore, the secondary power load is also higher. The quality of the power distribution system also has a greater impact on the overall grid. Fig. 4. 2. 7 shows a block diagram of the secondary power distribution using a Power Distribution Unit (PDU).

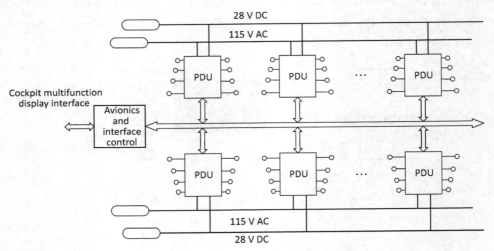

Fig. 4. 2. 7　Block diagram of secondary power distribution using power distribution

It can be seen from Fig. 4. 2. 7 that each group of secondary loads has one power distribution unit, which can be divided into n groups of power distribution units depending on the number of loads. Each grouped PDU can communicate with the avionics interface circuit for data communication.

As shown in Fig. 4. 2. 8, the hardware layout of the PDU consists of n AC and/or n DC power modules and control modules. The SSPC are used as switches to turn on and off each electrical load according to the commands from each Load Controller(LC).

Each module can exchange data with the data configuration control bus, and the control module receives commands to the aircraft interface communication bus.

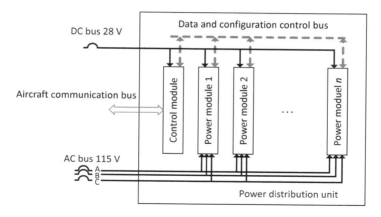

Fig. 4. 2. 8　Block diagram of the hardware layout of the PDU

3. DC Power Module Structure

Fig. 4. 2. 9 shows the DC power module structure diagram, the input is 28 V DC, after the input filter, respectively, to the solid-state power controller, which distributes power to the secondary Load 1, Load 2, Load 3, ⋯, Load k, respectively.

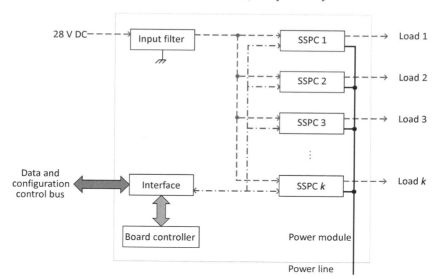

Fig. 4. 2. 9　DC power module structure

DC power module with board controller, interface circuit, etc. for data exchange with the configured control bus.

4. AC Power Module Structure

Fig. 4. 2. 10 shows the AC power module diagram. The three-phase 115 V AC current is transmitted to the respective solid-state power controller. SSPC is a non-contact electrical device composed of a semiconductor device, which has multiple functions of switching and breaking circuit, realizing circuit protection and transmitting state signal. The level of SSPC input control signal can be directly matched with the I/O interface of electrical and electronic controller EEC computer, which is suitable for the automatic distribution system. The automatic distribution system using SSPC is also called a solid state distribution system.

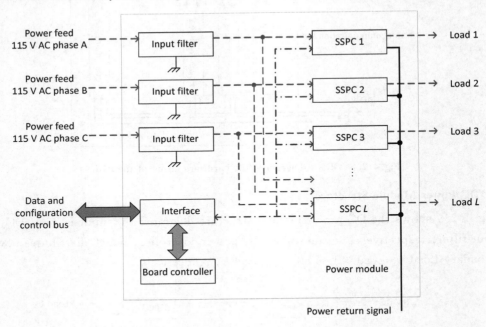

Fig. 4. 2. 10　AC power module structure

SSPC has no mechanical contacts, no mechanical moving parts, no sparks and arcs when connecting and disconnecting circuits, no mechanical failures, and long service life.

There are three types of SSPC on the aircraft according to the nature of power supply, namely 28 V low-voltage DC power system, 270 V high-voltage DC power system and 115 V AC power system SSPC. it should be noted that AC SSPC can be used for both 400 Hz constant frequency AC power supply and 360—800 Hz wide frequency AC power system.

The final stage power devices of DC solid-state power controllers mostly use low-voltage power field-effect transistors with fast response, low drive power, and low pass-state voltage. There are four typical types of DC SSPC loads, namely pure resistive loads, resistive inductive loads, resistive inductive and capacitive loads, and DC motor loads.

The AC SSPC is different from the DC SSPC in that the AC SSPC must achieve bi-directional current flow and blocking.

5. The Application of the SSPC on the B787

More than 900 SSPCs are used on the B787, whose reliability, no malfunction, small size, light weight, low loss and ease of use are the basic requirements. SSPC should also have line arc fault detection and protection capability. Integration, modularization, board and standardization has become the main development direction of SSPC. The EEC composed by computer and the remote control distribution neutral composed by SSPC is the

Line Replaceable Unit(LRU) of modern aircraft solid-state power distribution system.

When the contact current is less than 10 A, the 115 V AC and 28 V DC shorts utilizing the bi-metal principle have been replaced by solid-state power controllers consisting of power electronic circuits. Fig. 4. 2. 11 shows the airborne SSPC form factor.

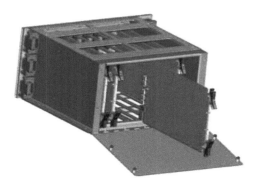

Fig. 4. 2. 11　Air board SSPC outline diagram

There are 17 remote power controllers RPDU on the B787 controlling a large number of SSPC. Most of the circuit breaker functions are controlled by remote control on the maintenance interface, usually using a data bus communicated by the common core system.

The circuit breaker must also have the function of arc extinguishing protection. As shown in Fig. 4. 2. 12, it is the functional equivalent of SSPC and ELCF.

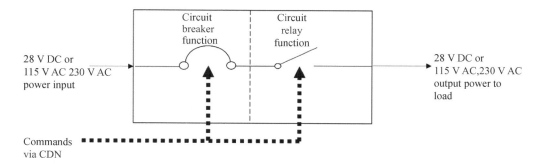

Fig. 4. 2. 12　Functional equivalence diagram of SSPC and ELCF

As shown in fig. 4. 2. 13, it is the ampere-second characteristics of traditional circuit breaker and solid-state power controller. Among them, the second characteristic error band of traditional circuit breaker is much wider than SSPC, and the accuracy of traditional circuit breaker is much lower.

One of the keys to the 270 V high-voltage DC power system that will be used in future advanced aircraft must be the use of no-contact, solid-state electrical appliances, and the SSPC is the core component of the 270 V DC system. In order to achieve uninterrupted power supply, both 270 V solid state power controller and 400 Hz AC solid state controller are required to be applied. Overseas 270 V and 28 V solid-state power controllers already

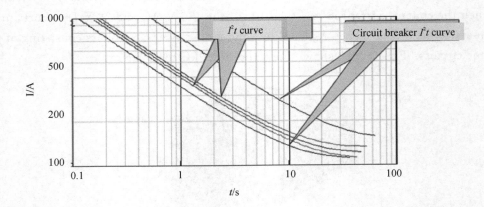

Fig. 4. 2. 13　I^2t characteristics of traditional circuit breakers and solid state power controller

have a series of products, and 270 V solid-state power controllers with rated currents of 50 A, 100 A and 130 A are under development. 130 A class SSPC has a cut-off current of 1 300 A. The key to SSPC is to reduce the transistor on-state voltage drop and volume size. The most important feature of the contact appliance is the small voltage drop and low loss when the contact is closed. Thus, no special cooling equipment is required. Due to the large on-state voltage drop of power electronics, the SSPC conducts with a large voltage drop and large losses, requiring the configuration of a heat sink, thus limiting the volume and weight reduction and limiting the development of the SSPC.

　　In addition to SSPC, 270 V distribution systems should also develop over current, differential current and short-circuit current detection and protection elements, arc detection and extinguishing elements, highly reliable connectors, and non-inductive DC bus bars.

4.3　Example of MEA Power Distribution

　　Typical more-electric aircraft are the A380 and B787 aircraft, which are used as an example for power distribution.

4.3.1　A380 Electrical System

　　The A380 is the first large civil aircraft of the modern era since some turboprop airliners in the early 1960s, reusing frequency conversion technology.

　　As shown in Fig. 4. 3. 1, the A380 electrical system components diagram, the AC power generation system consists of four 150 kV · A VF generators, whose operating frequency is 370—770 Hz, and the VF generators are highly reliable and efficient, but they cannot be connected in parallel and do not have the ability to supply power without interruption. Two 120 kV · A APU Constant Frequency Generators(CFGs) (400 Hz), four external power connections for ground. The A380 also has one 70 kV · A RAT generator for emergency AC

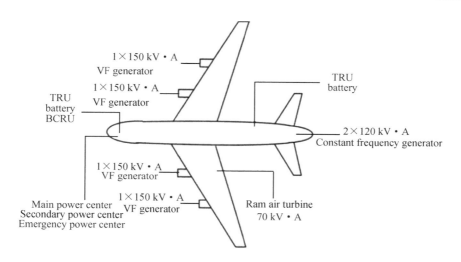

Fig. 4. 3. 1   A380 electrical system components

power generation.

The structure of the A380 AC PSS is shown in Fig. 4. 3. 2. Each 150 kV • A AC main generator is driven by the corresponding engine. Two APU generators are driven by their respective Auxiliary Power Units (APU). Each main generator supplies the corresponding AC junction bar (AC1—AC4) under the control of GCU (GCU 1—GCU 4). Each AC bar can also accept external power input for ground maintenance and support work. Because the aircraft generator is a frequency conversion, and the frequency of the AC power supply is related to the speed of the corresponding engine, the AC main junction bar can not work in parallel. The aircraft kitchen load, which forms a large part of the aircraft load, is scattered between the four AC confluence bars, as shown in Fig. 4. 3. 2.

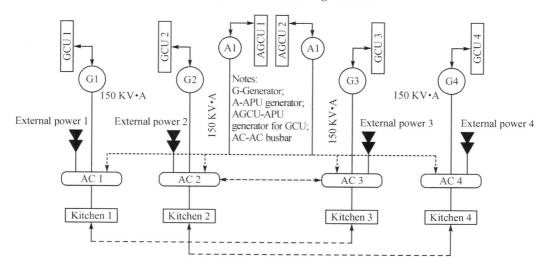

Fig. 4. 3. 2   A380 AC power system structure

1. A380 Electrical System Division

Fig. 4. 3. 3 shows that the A380 electrical system is divided into three channels, and that is E1, E2 and E3. The E1 channel is powered by AC1 and AC2, while E2 channel is powered by AC3 and AC4, and the RAT and static converters belong to E3 channel.

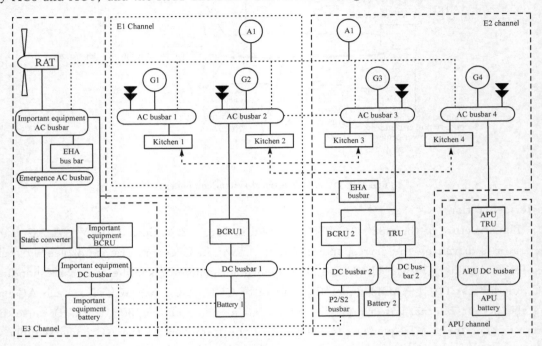

Fig. 4. 3. 3   A380 electrical system partition

Fig. 4. 3. 3 also shows that the E1 and E2 channels have two corresponding main generators, which are connected to the main generator with adjustable Battery Charge Regulator Unit(BCRU) and 50 A · h storage battery to obtain three practically independent power channels. That is, the E1 channel is AC1, AC2 + BCRU1 + battery 1; E2 channel is AC3 + BCRU2 + battery 2 and AC4; E3 channel is the emergency channel, it is composed of RAT + important equipment BCRU + important equipment battery.

2. A380 DC Power System

Fig. 4. 3. 4 shows the structure diagram of A380 DC power supply. It can be seen that the DC PSS has four 50 A · h batteries, three of which are charged by the BCRU, and the APU storage battery connected to the APU starter is charged by the Transformer Rectifier Unit (TRU) of the APU. The BCRU is generally composed of power electronic converter, which can realize constant current charging, constant voltage charging or floating charging to prolong the life of the battery.

In addition to supplying power to their respective loads (such as high-power kitchen loads), the AC confluence bar in Fig. 4. 3. 4 can supply power to the confluence bar of important equipment after BRCU conversion. Although the power on AC 1 to AC 4 is

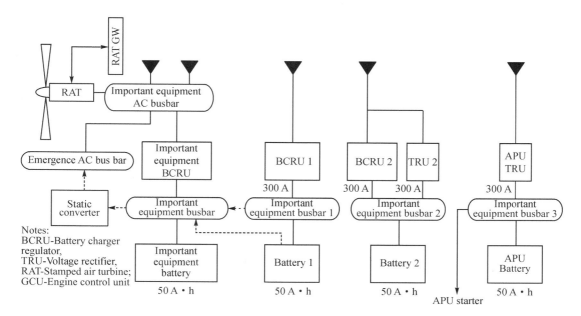

Fig. 4. 3. 4   A380 DC PSS structure

frequency conversion, it does not affect the realization of BCRU at all. BCRU converts the frequency conversion AC power supply of 115 V and 370—770 Hz to 28 V DC power supply with a capacity of 300 A.

AC bars of important equipment often supply power to high-power emergency loads, such as actuators of flight control system (EHA, EMA, EBHA, etc.). It is powered by three kinds of power supply, namely, under normal conditions by AC 1 and AC 4 power supply, when they both fail, by the emergency AC power supply from the RAT generator, if AC 1 and AC 4 and RAT all fail, then by the 50 A · h important equipment storage battery or 50 A · h battery 1 through the important equipment to the static converter, by the static converter to the emergency AC busbar power supply, thus giving important equipment AC busbar power supply.

The APU starter circuit is powered by a dedicated AC 4, and the starter is a DC motor. A 300 A APU variable voltage rectification unit TRU supplies the DC busbar 3 of important equipment and charges the APU accumulator battery, and when the AC 4 fails, the APU is started by the battery.

3. A380 Power Distribution System and Electrical System Control

The function of the power distribution system is to achieve power switching and protection. It mainly consists of one integrated Primary Electrical Power Distribute Central (PEPDC) and two aircraft load Secondary Electrical Power Distribute Central(SEPDC), which distribute electrical power to smaller electrical loads with power consumption less than 15 A/phase or less than 5 kV · A.

The six Secondary Electrical Power Distribution Busbar(SEPDB) distributes power to the cockpit and passenger comfort related life loads, which are not related to the flight safety of the aircraft and are usually of higher power, and distribute power as close to the loads as possible to reduce feeder cable weight. The domestic electrical loads include: galley equipment 120—240 kV · A, which is an intermittent load; refrigeration equipment about 90 kV · A, which is a continuous load; In-Flight Entertainment (IFE) about 50—60 kV · A, which consumes about 100 W of continuous electrical power load per seat; and cockpit lighting about 15 kV · A. The SSPC takes precedence over the circuit breaker for the secondary distribution.

The electrical system control of the A380 is controlled by a combination of dedicated units and integrated modular avionics and common processor input/output modules. Four main generators each have their own control units GCU 1—GCU 4; two APU generator control units AGCU 1 and AGCU 2; and one RAT generator control unit. They are used for the management of electrical loads, i. e. , control of load disconnections; and the monitoring of secondary loads, i. e. , monitoring the condition of the secondary distribution units. The monitoring of the circuit breakers is carried out where they are applied.

In any abnormal situation, the power distribution system is able to identify faults automatically and ensure that power is supplied to as many loads as possible and that any two generators are able to supply power to the entire electrical system. Since each generator is able to supply power to two AC busbars, each AC busbar is supplied in order of priority in the event that one generator fails.

4.3.2　B787 Power Distribution System

B787 distribution system is mainly composed of power supply system, electrical load, power distribution system and power distribution components, etc. The following is introduced.

1. Power Supply System

The top-level electrical system is shown in Fig. 4. 3. 5. It consists of 2 main engines and one APU engine, each with two starter/generators of 250 kV · A, so that each channel has a power of 500 kV · A.

The main feature is the use of three-phase 230/400 V inverter AC power supply, compared with 115/200 V, the voltage has increased by a factor of 1, which makes the feeder loss of the distribution system reduced, but there are higher insulation material requirements and the possibility of partial discharge "corona". The power supply frequency is 360—800 Hz, which helps to reduce the weight of the generator feeder, but the increase in frequency has an impact on the impedance and skin effect of the wire. The figure also shows two 225 kV · A APU-driven S/Gs. Each main generator is fed into its own 230 V AC main bus bar before being fed to the distribution system. The power supplies both feed 230 V AC loads and are converted to 115 V and 28 V supplies to feed conventional power-using loads.

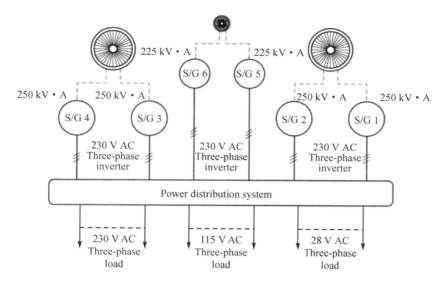

Fig. 4. 3. 5   Top-level electrical system

2. Electrical Load

As shown in Fig. 4. 3. 6, an electrical load diagram, there is no air delivery to the ECS, cockpit pressurisation system, wing anti-icing system, and other air pressure subsystems because the pilot air is no longer applied in the fuselage. The only induced air from the engine is the low-pressure fan air used for engine fairing anti-icing. Especially on modern engines, it is extremely wasteful of power to extract bleed air from the engine as the engine

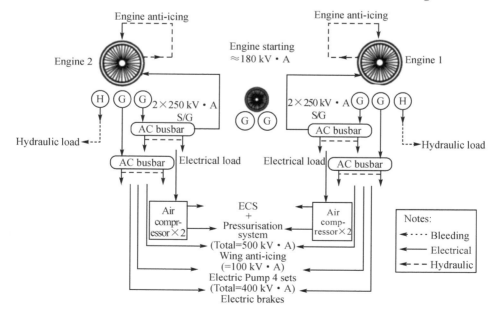

Fig. 4. 3. 6   Electrical load

pressure ratio and culvert ratio increase.

The main more-electric loads are environmental control systems and pressurisation, wing anti-icing, and electric pumps. Of these, the ECS and pressurisation eliminate the pilot air, which means that the air for the ECS and pressurisation systems needs to be electrically driven to generate pressurisation. On the B787, four large electrically driven pressurizers are required, extracting power in the range of 500 kV · A.

Instead of using induced air for wing anti-icing, electric heating pads embedded in the leading edge of the wing are used to provide anti-icing, which requires 100 kV · A of electrical power. Some of the engine drive hydraulic pumps are replaced by electric pumps. 4 new electric pumps require 100 kV · A each, for a total of 400 kV · A. If the "no jet engine" is used, the engines can not be started with high pressure air. The engines require 180 kV · A starter/generator to start the engines. The introduction of high power motors has important implications for the aircraft power distribution system.

3. Power Distribution System

A schematic diagram of the power distribution system is shown in Fig. 4. 3. 7. Primary power distribution is carried out by 4 main switchboards, 2 in the front electrical equipment bay and the other 2 in the rear electrical equipment bay. The rear switchboard also contains the motor controllers for the 4 Electric Motor Pumps (EMPs), 2 corresponding pumps housed in the engine hangers and 2 located in the mid-aircraft section. The motor controllers for the engine starter and the APU starter motor controller are also housed in the rear switchboard. The high power they have and the corresponding power losses generate a lot of heat, thus requiring liquid cooling of the main power distribution panel.

The electrically driven air conditioning assembly is located in the mid-section of the aircraft, and secondary power distribution can be achieved using RPDUs located at convenient locations on the aircraft. 21 RPDUs are located at the locations shown in Fig. 4. 3. 7(a).

Fig. 4. 3. 7(b) shows the relative positions of the main generator, the APU generator and the front and rear electrical equipment compartments. Since the main generator and APU generator are far away from the front and rear electrical equipment compartments, using 230/400 V AC voltage regime can reduce the weight of feeder lines. There are distribution boxes such as P100 and P200 in the equipment compartment, and six main power distribution boxes can control more than 200 medium and large electrical equipment.

(1) Aft E/E Bay Arrangement

Fig. 4. 3. 8 shows a schematic diagram of equipment placement in the Aft E/E bay arrangement, the diagram clearly shows the battery distribution box P49 and APU generator distribution box P150; left and right generator distribution boxes P100 and P200; left and

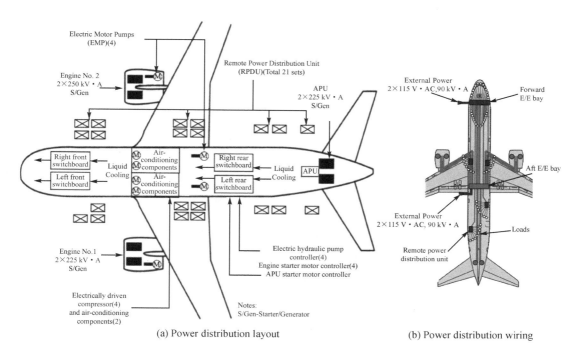

(a) Power distribution layout (b) Power distribution wiring

Notes: External power—external power supply; Aft E/E bay—rear electronic equipment bay; Remote power distribution unit—remote power distribution unit; Load—load.

Fig. 4. 3. 7 Schematic diagram of the power distribution system (B787)

right 270 V DC distribution boxes P700P and P800, these distribution units power system distribution equipment, close to the power generation unit, the use of wire or cable. The distance is close, which helps to reduce the weight of the cable.

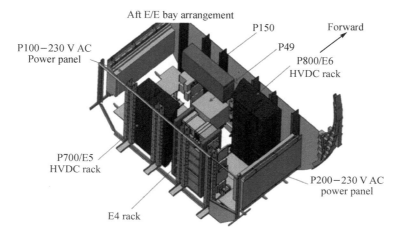

Fig. 4. 3. 8 Diagram of equipment placement in the rear electronics/electrical equipment bay

From Fig. 4. 3. 8, we can see that the components in the electronic equipment compartment are in the form of "shelves", which makes the placement of equipment clear and easy to find and troubleshoot, and once a failure occurs, the replacement unit is replaced on site, which is fast and efficient.

(2) Former E/E Bay Arrangement

As shown in Fig. 4.3.9, it is a diagram of equipment placement in the front electronic/ electrical equipment compartment.

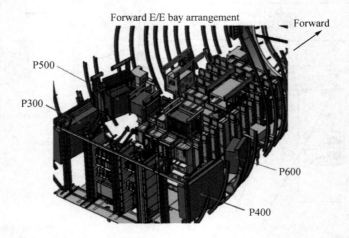

Fig. 4.3.9   Schematic diagram of equipment placement in the front electronics/electrical equipment bay

As can be seen in Fig. 4.3.9, the distribution panels are the transformer rectifier unit P300, 115 V AC distribution box P400, left and right 28 V DC distribution panels P500 and P600, installed in the front electronic/electrical equipment compartment, where there are more power-using equipment. These distribution, also known as secondary distribution, are transformed by the main power supply for distribution. As shown in Fig. 4.3.10, it is the distribution system schematic block diagram.

4. Power Distribution Components

The aforementioned distribution components of the inverter AC power supply, including switches, relays, contactors and connectors, are the same as the 400 Hz constant frequency AC PSS, and this inheritance reduces the risk of the development of the inverter AC PSS. Tab. 4.3.1 shows the list of busbars and switching appliances in each distribution box.

GNR in Tab. 4.3.1 is the relay for disconnecting the motor midpoint from the ground wire during the starter/generator operation. L1SC and L2SC are the connection circuit breakers between starter/generator 1 and 2 and the power electronic converter during starter operation. The contactors L1CC, L2CC, L1C1, R1C1, R1CC, R2CC, R1C2, L2C2 and NGC in the P700 and P800 distribution boxes are contactors for switching on or disconnecting the high-power speed controller and the motor. R1C2 and NGC are contactors used to turn on or off between the high power speed controller and the motor.

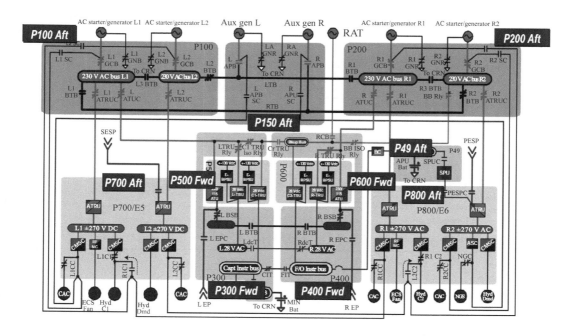

Fig. 4. 3. 10 Block diagram of a power distribution system

Tab. 4. 3. 1 B787 main distribution box devices

SN	Name	Main equipment	Remarks
1	Left generator distribution box P100	230 V AC bus L1, 230 V AC bus L2, L1GCB, L1GNR, L1BTB, L1ATRUC, L1SC, L2GCB, L2GNR, L2BTB, L3BTB, L2ATRUC, L2SC	
2	Right generator distribution box P200	230 V AC bus R1, 230 V AC bus R2, R1GCB, R1GNR, R1BTB, R1ATRUC, R1SC, R2GCB, R2GNR, R2BTB, R3BTB, R2ATRUC, R2SC	Inside the aft electronics bay
3	APU generator distribution box P150	RAPB, RAGNR, RAPUSC, LAPB, LAGNR, LAPUSC	
4	APU battery distribution box P49	APU, BAT SPU, SPUC	
5	Left 270 V distribution box P700	L1±270 V DC bus, L2±270 V DC bus, ATRU1, ATRU2, CMSC 4 units, RFMC 1 unit, SESPC, L1CC, L2CC, L1C1, R1C1	
6	Right 270 V distribution box P800	R1±270 V DC bus, R2±270 V DC bus, ATRU 3, ATRU 4, CMSC 4 unit, RFMC 1 unit, SESPC, R1CC, R2CC, R1C2, L2C2 and NGC	

(**continued**)

SN	Name	Main equipment	Remarks
7	Left 115 V/28 V distribution box P300	L 115 V AC bus, L 28 V DC bus, Capt Instr bus, L EPC, MBR, CIT, LDCT, LBTB, LBSB, L TRU RLY, C1 TRU ISO RLY	Inside the former electronics bay
8	Right 115 V/28 V distribution box P400	R 115 V AC bus, R 28 V DC bus, F/O Instr bus BKUP bus, REPC, RCB, R TRU RLY, FIT, RDCT, RBTB, R BSB, C1 TRU RLY, BB ISO RLY	
9	Left secondary power bank P500	ATU 1 unit, TRU 2 units, BPSU 2 units, two±130 V DC buses	
10	Right secondary power supply box P600	ATU 1 unit, TRU 2 units, BPSU 2 units, two±130 V DC buses	

4.4 Summary

The power distribution system is an important link between power supply and power-using equipment, and is the basis for achieving fault tolerance and uninterrupted power supply. It is impossible to eliminate the fault of power supply, grid conductor and power-using equipment, and the key is that the power supply should still be reliable to power-using equipment in case of fault.

Loads are classified as critical, primary and secondary according to their importance. Critical loads are those with the minimum power supply necessary for safe flight and landing.

The fundamental change in the load characteristics of MEA is due to the continuous equipping of power electronics in aircraft and the massive use of power electronic converters and electric motors, which has led to important changes in the electrical equipment of aircraft. Many technological advances have been made in the generation, switching and protection of electrical power, which are beginning to have an impact on classical electrical systems, with new devices, materials and theories emerging.

The ELMC's built-in electrical and electronic controller, the EEU, enables detection and switching of internal busbars; automatic load management; distribution network fault detection and protection to improve the performance and maintenance of the aircraft power system; and reporting of fault handling and power consumption to the power system handler, reporting of fault handling and power usage to the PSS processor.

More than 900 SSPCs are used on B787, whose reliability, non-misoperation, small size, light weight, low loss and ease of use are the basic requirements. SSPC should also have line arc fault detection and protection capability. Integration, modularization, board and standardization have become the main development direction of SSPC. The EEC composed by computer and the remote control distribution neutral composed by SSPC is the basic replaceable unit LRU of modern aircraft solid-state power distribution system.

4.5 Exercises

1. How many categories is the load of aircraft divided into?

2. What are the characteristics of the electrical load of an MEA?

3. What are the requirements for the aircraft power grid?

4. What is the difference between the power distribution mode of traditional aircraft and that of MEA?

5. What is the impact of the secondary power distribution of conventional aircraft and MEA on the cockpit?

Chapter 5 Electric Actuating Technology of MEA

5.1 Introduction

The aerospace field is undergoing a transition from the use of mechanical energy, hydraulic energy, and pneumatic energy to the use of electricity, thereby realizing an all-electric aircraft that only uses electrical energy. For example, electric motors can convert electrical energy into mechanical energy to drive actuators, various pumps, compressors, and other systems that need to change speed. If the advanced power electronic technology is combined with the control strategy, the electric drive device can achieve an increase in overall efficiency, a reduction in weight and a reduction in cost, which meet the requirements for reliability. Therefore, the ultimate goal of aircraft design and manufacturing is to achieve an AEA that only uses electrical energy. It is estimated that AEA can reduce the weight by about 10% and reduce the fuel consumption by about 9%.

The concept of electric aircraft was first proposed in 1916. During World War II, British "V" bombers used electricity for main flight control and other drive functions. As mechanical, hydraulic and pneumatic drive technologies are mature and widely accepted, secondary power drivers are mainly used in high-power drive situations, such as hydraulic devices for landing gear, braking systems, flight control systems; pneumatic devices for pressurisation, deicing and air conditioning equipment; and electric power devices used for avionics and electrical equipment.

Since aircraft safety is the most important basic requirement, conservative methods are often adopted when implementing new design ideas or technologies. Consequently, the use of electric control and drive technology has been showing a gradual increase in the aerospace field. Mechanical actuators are gradually being replaced by electro-hydraulic drives with electronic servo valve control devices.

The A380's EHA provides partial hydraulic drive pumps and accumulators, allowing it to work with electrical energy. It simplifies the mechanical linkage mechanism and reduces the number of hydraulic oil supply lines, thereby reducing weight and simplifying maintenance. The fuel pump of the more-electric engine replaces the hydraulic pump, providing flexibility in system efficiency, weight, size, and speed control.

At the end of the 1970s, mono-phase power appeared in the secondary power supply, but there is no obvious progress had been made at that time. In the 1980s, the United States started developing electric drive technology for civil and military aircraft, that is, the use of permanent magnet materials, power semiconductor materials and technological advancements in control strategies make the electric drive show advantages in terms of safety and reliability. Those advantages have greatly promoted the progress of the aviation industry. In the 1990s, the US Air Force launched a research project called "MEA" to promote electric drive technology gradually. Tab. 5. 1. 1 shows the applications of MEA motor drives.

Tab. 5. 1. 1　Application of MEA motor driver on airplane

SN	Subsystem name	Motor drive quantity/unit	Large motor power/kW	Total power/kW
1	Flight control	28	50	80
2	Environmental contol	10	10	40
3	Engine starter/generator system	6	125	125/channel
4	Landing system	20	5	30
5	Fuel pump	10	9	35
6	Air pressure system	2	15	30
7	Hybrid system	10	1	20

In recent years, a lot of efforts have been made in electric drive, and MEA have attracted more and more attention. Electric drive systems must meet high reliability requirements and have airworthiness, etc. , so the progress is quite slow.

There are fuel delivery pipelines, hydraulic pipelines, and air pressure pipelines on the aircraft, all of which must use valves to control the degree of opening to achieve flow regulation. The cockpit temperature adjustment valve is adjusted by controlling the hot and cold air flow, and it also needs the control of the electric motor. The electric actuation on the MEA requires a large number of electric mechanisms, and its importance is also increasing, which will be introduced separately below. Electric motors are mainly used in the following areas:

① Linear actuators: adjustment of actuators for engine control and flight control systems.

② Rotary actuator: actuation of electrical position, actuation of flaps and slats through roller screws.

③ Valve actuation: control the valve opening degree of fuel, hydraulic, air and auxiliary systems.

④ Engine start: start the engine, APU and other equipment to a self-sustaining operating state.

⑤ Various pumps: fuel supply pumps, hydraulic pumps, recovery pumps, valves for air and auxiliary system control.

⑥ Gyroscope: power provided for gyroscope flight instruments and automatic flight instruments.

⑦ Fan: an electric fan that heats the cabin electrically and a cooling fan that provides air to passengers.

Fuel pump motors, hydraulic pump drive motors, and fan motors exist on every aircraft, but their number and power level are different. Tab. 5. 1. 1 lists the typical electricity consumption of MEA. It can be seen from the table that the proportion and power of electric motors are relatively large among the electrical equipment. Electric motors play an important role in MEA already.

5.2 Application of Electric Actuation Technology

Since electric power has the characteristics of clean, quiet, and easy to achieve automatic control, aircraft are increasingly using electricity to work. The main applications in electric actuation are: fly-by-wire actuators, EHA and EMA, etc.

5.2.1 Fly-by-Wire Actuator

Fly-by-wire flight control systems appear on civil aircraft (such as A320), making the connection between flight control systems and fly-by-wire devices more convenient. The first generation of aircraft using the technology can work in 3 different modes: full electric mode, direct electrical link mode, and mechanical recovery mode. The all-electric mode adopts the calculation and protection of all fly-by-wire, which is already a normal operating mode. The direct electrical link mode usually provides basic calculations or may only have the ability to transmit direct electric signals when the main fly-by-wire mode cannot be used. The mechanical recovery mode provides the basic (mechanical) ways to make the aircraft fly, and only a limited number of flight control surfaces can be used for control after the failure of the fly-by-wire and direct electrical link modes.

As shown in Fig. 5. 2. 1, the block diagram illustrates the interface when the Actuator Control Electronic(ACE) device to realize the connection with the actuator. The electrical circuit is closed around the control circuit of the actuator instead of a mechanical closed circuit. After the control instructions being processed by the digital fly-by-wire instructions of the flight control system or direct electrical link instructions being processed by ACE, the actuator servo valve is provided analog instructions. In this way, the hydraulic power of the aircraft system is supplied to the corresponding side of the piston of the actuator cylinder, and the piston is moved to the required position. Piston position detection uses a Linear Variable Differential Transformer (LVDT) detection, and feeds back the position signal to ACE, forming a closed actuator loop.

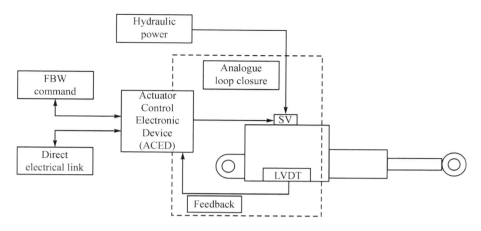

Fig. 5. 2. 1   Schematic diagram of fly-by-wire actuator

5.2.2   Electro-Hydraulic Actuator

The EHA is realized by using advanced power electronic device and control technology to provide more effective flight control action for the flight control system.

Fig. 5. 2. 2 shows the schematic diagram of the EHA. The EHA can only extract more energy when there is a manipulation command. Unlike the aforementioned actuators, in most flights, the actuator is commanded for a short period of time and the energy consumption is large, resulting in more energy extraction from the engine and more fuel consumption. In EHA, the actuators are stationary during flight without control commands. The EHA achieves this by feeding three-phase AC power into the power drive device, which in turn drives a variable speed motor and a constant flow hydraulic pump. Constructing a local hydraulic system to the actuator in a similar form to the IAP, the extracted power only needs to keep the electronic controls in standby state when there is no command. When there are instructions from the ACE, the power-drive electronic device can drive the variable speed motor quickly enough. Then the actuators are pressurized enough to

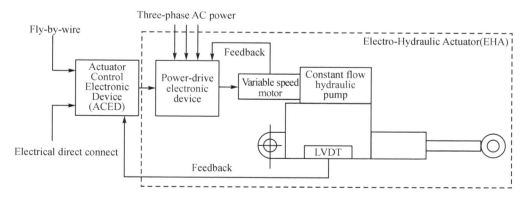

Fig. 5. 2. 2   Schematic diagram of EHA

move the corresponding control surface as instructed. Once the instruction requirements are met, the power electronics return to a normal standby state. So power is drawn from the power bus only when the actuator is running, which saves the energy considerably.

Due to the development of power electronics technology, EHA is used in various aircraft, including A380 and F－35. If a three-phase AC 115 V/400 Hz power supply (A380) is used, the actuator uses a three-phase bridge rectifier circuit and filters to form 270 V DC, which is used to drive a 270 V DC brushless motor and a constant flow pump. If there is a 270 V DC power supply (F－35) on the aircraft, the rectification and filtering parts can be omitted, thereby improving the efficiency, and reducing the volume and the weight.

These actuators all use microprocessors to improve control technology and performance, introduce digital control technology, and can also communicate directly with Flight Control Computers(FCCs) by considering (ARINC429/ARINC629/1553B). The development of EHA and EMA uses more-electric power, especially when the aircraft is equipped with 270 V, which is closer to the design concept of AEA.

5.2.3　Electric Mechanical Actuator

The EMA replaces the electric signal actuation and power actuation of the EHA, and uses the motor and the reducer assembly to generate the driving force to move the actuator. EMA has been used in aircraft for a long time as a trim and door action, mainly because the power, thrust and response time are very low and short, which is not conducive to the requirements of flight control action.

In recent years, the progress of following three key technologies, the development of 270 V DC rare earth permanent magnet motor, the development of high-power solid-state switching devices and the development of microprocessors, make it possible to be used in flight control.

As shown in Fig. 5.2.3 it is the schematic structure diagram of EMA. EMA is a more-electric type of screw actuator. The principle of EMA is the same as that of screw actuator, and the difference is that it applies power-drive electronic device to drive the DC brushless

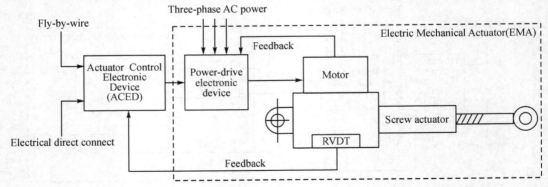

Fig. 5.2.3　Schematic diagram of EMA

motor and drag a reduction gearbox to generate rotational motion, so that the screw actuator can be extended or retracted to meet the input order. Therefore, EMA is used to driving the horizontal stabilizer, flaps and slats on the aircraft, and it is also used in the helicopter flight control system. However, EMA should pay attention to the stuck problem in use, which limits its application in the main flight control system.

Fig. 5. 2. 4 shows an advanced electric drive EMA, which adopts mechanisms such as electric motors, roller screws, and gears.

Fig. 5. 2. 4   Advanced electric drive EMA

The motor in the Electro-Mechanical Actuator (EMA) mechanism or the Electro-Hydrostatic Actuator (EHA) mechanism is a servo motor. In addition to the requirement of high power density, it should have good dynamic performance and a wide range of speed regulation.

5.2.4 Typical Applications of Flight Control Actuators

Due to different application scenarios, flight control actuators have different power, power supply and types. Tab. 5. 2. 1 lists the typical applications of actuators on civil aircraft.

Tab. 5. 2. 1   Typical applications of flight control actuators

SN	Actuator type	Power source	Typical application			
			Main flight controls	Spoiler	Horizontal stabilizer	Flaps & slats
1	Conventional linear actuator	Aircraft hydraulic system B/Y/G or L/C/R[(1)]	√	√	—	—
2	Conventional screw actuator	Aircraft hydraulic or electrical systems[(2)]	—	—	√	√
3	Integrated Actuator Package(IAP)	Aircraft electrical system 115 V AC	√	√	—	—
4	Electric signal hydraulic actuator	Aircraft hydraulic system	√	√	—	—

(Continued)

SN	Actuator type	Power source	Typical application			
			Main flight controls	Spoiler	Horizontal stabilizer	Flaps & slats
5	Electro-Hydraulic Actuator(EHA)	Aircraft electrical system[3][4]	√	√	—	—
6	EMA	Aircraft electrical system[3]	—	—	√	√

Remark: (1) B/Y/G=blue/yellow/green or L/C/R=left/center/right.

(2) For horizontal stabilizers, flaps and slats, both hydraulic and electrical power are often applied for redundancy.

(3) Three-phase 115 V AC is converted into 270 V DC on civil aircraft.

(4) Aircraft electrical system using 270 V DC on F – 35/JSF.

5.2.5 Flight Control Actuation Electrical System

The A380 aircraft weighs about 600 tons, and its most prominent feature and difficulty is the high power of its flight control actuation device, so it is introduced as an example.

Fig. 5.2.5 shows the schematic diagram of the hydraulic and electrical power generation of the A380 aircraft. The main power source is composed of electrical system 1 (E1) and electrical system 2 (E2) formed by 4 Alternating Current (AC) generators and an emergency system (E3) composed of RAT generators. The hydraulic system consists of the green system on the left and the yellow system on the right.

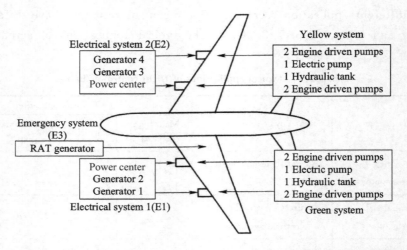

Fig. 5.2.5 Schematic diagram of A380 hydraulic and electrical power generation

The A380 flight control actuator configuration is listed in Tab. 5.2.2. Many actuators are supplied pressure only by the aircraft's green (G) hydraulic system (left system, driven by engines 1 and 2) and yellow (Y) hydraulic system (right system, driven by engines 3 and 4).

Tab. 5. 2. 2　A380 flight control actuator matrix table

Left wing			Right wing		
Aileron	Spoiler		Spoiler		Aileron
Inside G	1	Y	Y	1	G Inside
AC E2	—	—	—	—	AC E2
Middle Y	2	G	G	2	Y Middle
AC E1	—	—	—	—	AC E1
Outside Y	3	Y	Y	3	Y Outside
G	—	—	—	—	G
—	4	G	G	4	—
—	5	Y+AC 2E	Y+AC 2E	5	—
—	6	G+AC 1E	G+AC 1E	6	—
—	7	Y	Y	7	—
—	8	G	G	8	—

Right elevator	Horizontal stabilizer		Left elevator
Inside AC E1	G		AC E1 Inside
G	Y		Y
Outside AC E2	AC E2		AC E2 Outside
G	Rudder (up)	Rudder (down)	Y
—	1Y+AC E1	1G+AC E1	—
—	2G+AC E2	2Y+AC E3	—
—	—		AC E1 AC1
—	—		Important equipment channel
—	—		Side 1
—	—		AC E2 AC2
—	—		Important equipment channel
—	—		Side 2
—	—		AC E2 AC2
—	—		Important equipment channel (RAT)

Tab. 5. 2. 3 lists the aircraft actuating mechanism condition of A380 aircraft. A380 has 7 kinds of flight control rudder surface actuating mechanisms, including 28 hydraulic actuators and 19 electric mechanisms, of which 8 are EHAs mechanisms, 8 are electric backup hydraulic actuating mechanisms EBHA, and 3 are electric-electric actuating mechanisms.

Tab. 5. 2. 3　Aircraft actuation mechanism status

SN	Name	Number (RSAM)	Number (HA)	Number(EA)
1	Aileron	12	8	4(EHA)
2	Elevator	8	4	4(EHA)
3	Rudder	2	—	4(EBHA)
4	Horizontal stabilizer	1	2	1(EMA)
5	Spoiler	16	12	4(EBHA)
6	Flap	1	1	1(EMA)
7	Slat	1	1	1(EMA)

Fig. 5. 2. 6 shows the schematic diagram of the A380 flight control actuation system. It can be seen from Fig. 5. 2. 6 that many positions are actuated by a combination of conventional hydraulic actuators (H) and EHA.

① Each wing consists of 6 outboard aileron rudder surfaces and 8 spoiler rudder surfaces, actuated by conventional hydraulic actuators, namely yellow and green systems.

② Intermediate and inboard aileron surfaces, and inboard and outboard elevator surfaces are actuated by both hydraulic actuators (H) and EHA, each of which acts as the other actuator in case of failure.

③ Two spoiler surfaces (No. 5 and No. 6 on each wing) and two rudders are actuated by Electric Backup Hydraulic Actuators (EBHA), which combine the characteristics of hydraulic actuators (H) and EHA.

④ The horizontal stabilizer actuator is independently driven by the green/yellow hydraulic channels and E2.

As can be seen from Fig. 5. 2. 6, there is also the drive of flaps and slats. The slats are driven by the green system or E1; the flaps are driven by the green or yellow channel. The EBHA receives input from the relevant hydraulic channel (green or yellow channel) and electrical channel (E1 or E2 or E3 AC important equipment channel (RAT) in special cases). As far as the rudder is concerned, the upper surface is driven by green and yellow hydraulic channels, E1, E2 and AC2; the lower surface is driven by green and yellow channels, E1 and E3.

EBHA can have two operating modes, namely hydraulic mode and standby EHA mode. Fig. 5. 2. 7 shows the operating mode of A380 electric backup hydraulic actuator.

When operating in hydraulic mode, the actuator receives the power of the relevant green or yellow hydraulic system, and the servo valve adjusts the pressure of the input actuator according to the instructions of the fly-by-wire computer. When operating in standby mode, the actuator works the same as the EHA. Accepting electrical power from the aircraft's AC electrical system, the fly-by-wire computer feeds commands into the EHA control assembly. The direction and speed of the motor determine the direction and speed of movement of the actuator piston.

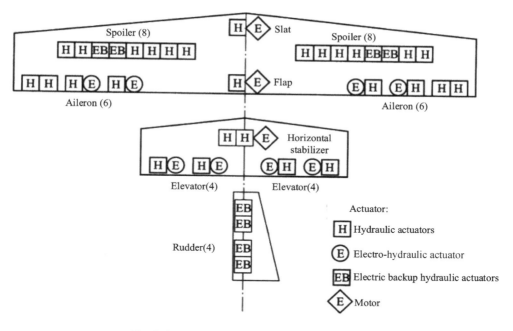

Fig. 5. 2. 6   A380 flight control actuation system

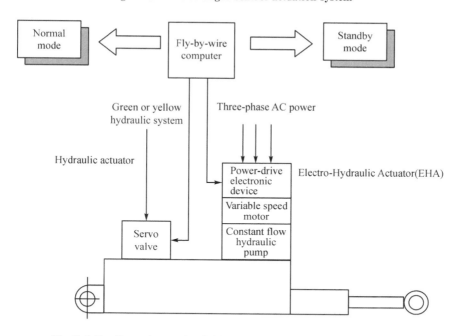

Fig. 5. 2. 7   Operating mode of A380 electric backup hydraulic actuator

5.2.6   Introduction to Motor Drive Technology

As shown in Fig. 5. 2. 8, it is the motor power drive mode. The drive used for the motor often passes through the power converter, which usually has direct power conversion and one stage power conversion. They are different in the design and composition of the driving circuit of the motor.

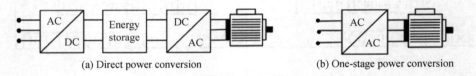

(a) Direct power conversion   (b) One-stage power conversion

Fig. 5. 2. 8   Motor power drive mode

The motor driving device is constructed by using power electronic technology, which effectively solves the driving problem of the motor. As shown in Fig. 5. 2. 9, it is the structure diagram of an EHA. The power supply is phase voltage 115/200 V, 360—800 Hz variable frequency AC power supply. The DC bus voltage is formed through the input filter, rectifier and filter circuit, typically at 270 V. This voltage passes through the inverter to form the drive power to drive the Permanent Magnet Synchronous Motor (PMSM). The PMSM drives the hydraulic pump and then drives the flight surface.

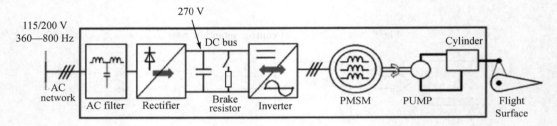

Fig. 5. 2. 9   Structure diagram of EHA

As shown in Fig. 5. 2. 10, the common inverter six-bridge power drive circuit is used to drive the PMSM. Its power supply voltage value is 270 V DC, and then passes through the inverter to form the drive power of the PMSM.

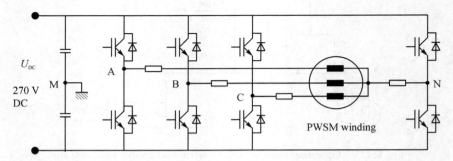

Fig. 5. 2. 10   Power device connection of four-bridge inverter

If a motor powered by a DC power supply is defined as a BLDCM in a broad sense, an asynchronous motor, a synchronous motor, a switched reluctance motor, a double salient motor and a stepping motor can all constitute a BLDCM.

Besides the heavy use of electric motors in typical actuation systems and environmental control systems, electric motors are also required in applications such as engine fuel supply and starting, door opening and closing, and aircraft landing gear control, which will not be described in detail here.

Fig. 5. 2. 11 shows the feeding diagram of the high-power motor of the MEA(B787).

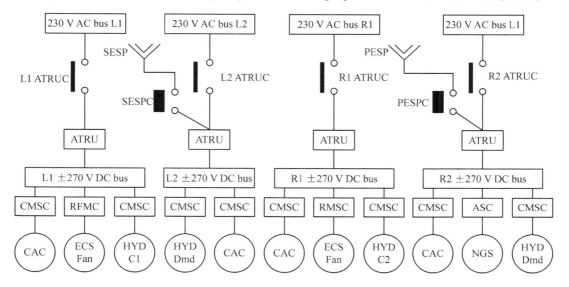

Fig. 5. 2. 11   The feeding diagram of the high-power motor of the MEA(B787)

The power supply for the high-power motor on the MEA/AEA represented by B787 mainly has 230 V AC and 270 V DC generated by Auto Transformer Rectification Unit (ATRU), it can be seen from the figure that four 270 V DC power supplies can provide power for various high-power motor equipment.

Fig. 5. 2. 11 shows the 270 V electrical loads, such as the Cabin Air Compressor (CAC), Environmental Control System(ECS) fan, the Hydraulic(HYD) Drive motor device (Dmd), and the Nitrogen Generation System(NGS). Those are all high-power electrical loads driven by electric motors.

5.3 Permanent Magnet Generator and Power Converter for Flight Control

5.3.1 Introduction to Rare Earth Permanent Magnet Materials

Rare earth permanent magnet materials have high BH_{max}, linear magnetic curve and low magnetic permeability, so rare earth permanent magnet motors have the characteristics of simple structure, small size, light weight, low loss and high efficiency, which are valued by the aviation industry and are widely used.

Rare earth permanent magnet materials are permanent magnet materials born in the

1960s. There are mainly samarium cobalt permanent magnets and rubidium iron boron permanent magnets. The latter cannot be used in aviation due to poor temperature characteristics. There are two types of samarium cobalt permanent magnet, namely RCo_5 (Type 1:5) and R_2Co_{17} (Type 2:17). The magnetic performance of the R_2Co_{17} is better than that of the RCo_5, and the BH_{max} of the permanent magnet can reach 258 kJ/m^3, remanence $B_r = 0.85—1.15$ T, Coercivity $H_c = 480—800$ kA/m. The demagnetization curve is a straight line and the temperature coefficient of B_r is -0.03%, which is relatively low. The Curie temperature is 720 ℃ to 880 ℃.

Rare earth permanent magnet Brushless DC Motor(BLDCM) usually refers to a BLDCM composed of a PMSM, a motor rotor position sensor, a power converter and a digital controller. This motor has a linear torque-current characteristic, that is, the current output torque is proportional to the armature current. It has a strong overload capacity, generally the maximum torque can reach 3 times the rated torque, and the maximum torque can reach 8—10 times. In the constant torque speed regulation area, the output power of the motor with a certain output torque is proportional to the speed, so when the power supply voltage is constant, the input current of the motor is proportional to the motor speed. In the condition that the motor speed and torque are constant, the input current of the motor decreases when the power supply voltage increases, and the power is constant.

Rare earth permanent magnet BLDCM has two operation modes, 120 degree square wave current operating mode and sine wave current operating mode, the latter has smaller torque ripple, and the application of magnetic field direction control can make it work smoothly at very low speed (for example, 0.1—0.01 r/min), and has a larger speed regulation range, and the speed regulation is above 10 000:1.

The disadvantage of rare earth PMSM is that it is difficult to increase the speed with field weakening. In order to further expand the high-speed operating area of the motor, many scholars have proposed a field weakening method of the PMSM. Using a hybrid excitation structure is one of the effective methods.

Aircraft electric ECS requires high-speed motor and an important subject of high-speed motor is to use rotor suspension technology. At present, the long-term rotating speed of the 150 kV · A synchronous generator lubricated with lubricating oil has reached more than 40 000 r/min. The air suspension bearing has been used, the magnetic suspension technology has been quite mature, and the no bear motor technology has entered the stage of industrial application, all of which have created conditions for the application of high-speed motors in more-electric and AEA.

It must be noted that high-speed and ultra-high-speed motors are inseparable from power electronic conversion technology. It is impossible to be directly powered by the industrial frequency power supply (such as the mains power supply of 50 Hz or 60 Hz), and it is impossible to directly powered from the 400 Hz aircraft AC power supply on the plane. Therefore, this type of motor is still essentially a BLDCM. Due to the simple and strong rotor structure, the motor can work at a very high speed when high-strength magnetic materials are used.

5.3.2 Permanent Magnet Generator

In the 1970s, the third-generation of America fighter F – 16 was born, and the use of the aircraft's fly-by-wire flight control system significantly improved the tactical and technical performance of the aircraft. As shown in Fig. 5. 3. 1, the F – 16A/B four redundancy flight control power supply is used. In order to improve the reliability of the flight control system, the flight control computer(FCC) adopts the control scheme of more than four redundancy from A to D, Therefore, 4 independent 26 V, 800 Hz power supplies No. 1 FCS—No. 4 FCS are required to supply power to the flight control computer A—D respectively.

In order to improve the reliability of the flight control power supply, each flight control power supply has 4 power supply channels, one of which is powered by dedicated batteries (4 sets, not marked in Fig. 5. 3. 1) for flight control, the other is powered by dedicated permanent magnet generators, the third is powered by No. 1 28 V DC bus on the aircraft, and the forth is powered by No. 2 28 V DC bus.

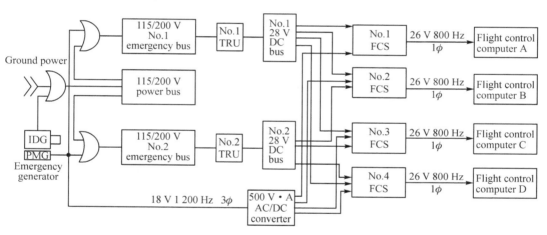

Fig. 5. 3. 1 F – 16A/B 4 degree redundant flight control power supply

It can be seen from Fig. 5. 3. 1 that the permanent magnet generator for flight control generates 18 V, 1 200 Hz three-phase alternating current, and generates 4 independent 28 V DC through the AC/DC converter. The 28 V DC offer power to the 4 degree redundant Flying Control System Power Supply (FCSPS), which can supply power only after the Emergency Power Unit(EPU)operation.

1. The EPU Operates in Two Ways

① Utilize the engine compressed air at the 13th stage to drive. As long as the engine works normally and there is bleed air, the EPU can work.

② When the engine fails, the gas-driven turbine is driven by the combustion of the unit fuel in the combustion chamber. Due to the limited fuel, the EPU can only work for a short time.

2. Special Power Supply for Redundancy Control

When the EPU is operating, the emergency alternator generates electricity. Even if the IDG fails, the 115 V/200 V emergency buses of No. 1 and No. 2 still have power, and the TRU(Transformer Rectifier Unit) can still supply 28 V to No. 1 and No. 2. DC buses. In this emergency situation, as long as the No. 1 and No. 2 TRUs and the corresponding 28 V DC buses have no failure, the 4-degree redundant flight control power supply will still supply power for multiple channels. Only when the No. 1 and No. 2 28 V DC buses also lose power, the flight control power supply is powered by the dedicated permanent magnet generator in the emergency generator. It can be seen that the permanent magnet generator plays a very important role.

(1) Four Redundancy Special Power Supply for F – 16C

The F – 16C entered service in 1978. In order to further improve the reliability of the aircraft power supply and the special power supply for flight control, a 10 kV · A VSCF (Variable Speed and Constant Frequency) AC power supply is installed on the F – 16C aircraft. There are 5 permanent magnet generators in those three-stage brushless alternators, one of which is used as an auxiliary exciter for the main generator, and the other four are dedicated to supply power to the flight control power supply. Under normal circumstances, it mainly supplies power to the flight control power supply, as shown in Fig. 5. 3. 2.

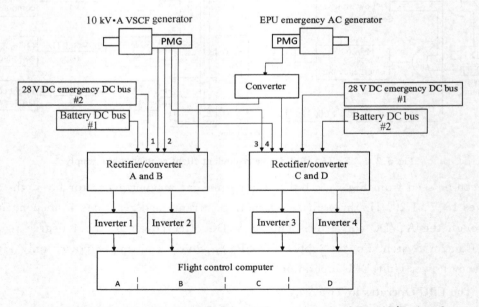

Fig. 5. 3. 2 Schematic diagram of flight control power supply of F – 16C/D aircraft

The dedicated power supply for the flight control is a 4-degree redundant power supply, which consists of a rectifier/converter and an inverter. 4 rectifiers/converters are placed in the left and right chassis, and each of the 4 inverters has a chassis. Each rectifier/converter is powered by four power supplies, which are:

① PMG(Permanent Magnet Generator) of VSCF power supply.

② PMG and converter of emergency alternator.

③ 28 V DC emergency DC bus.

④ Battery DC bus.

The emergency alternator is powered only when it is operating, not when the mains are operating normally. There is also a dedicated battery in the dedicated power supply for the flight control.

The 4-PMG dedicated to flight control in the VSCF generator will supply power to the flight control power supply as long as the engine works normally, so these 4 PMGs become the main power supply of the flight control power supply.

(2) Three Redundancy Special Power Supply of B777

Fig. 5. 3. 3 shows the block diagram of the flight control power supply of the B777 aircraft. The flight control power supply is a 3-degree power supply, which are the left, center and right flight control power supplies. It can be seen from the figure.

① The left flight control power supply has 3 channels of power supply: the left 28 V DC bus, the battery bus and the dedicated PMG for flight control inside the VSCF power generator.

② The center flight control power supply has 4 channels of power supply: the 28 V DC instrument bus of the pilot, the direct bus of the battery, the No. 2 PMG in the left VSCF power generator, and the No. 1 PMG in the right VSCF power generator.

③ The right flight control power supply has 4 channels of power supply: the right 28 V DC bus and the No. 2 PMG in the right VSCF power generator. There are also dedicated batteries LB, CB and RB in the three flight control power supplies.

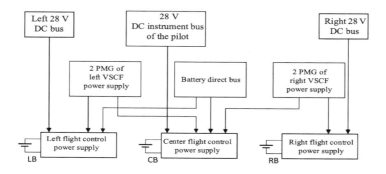

Fig. 5. 3. 3   Block diagram of the flight control power supply of the B777 aircraft

When the engine is operating normally, the flight control power supply is powered mainly by the permanent magnet generator in the VSCF power supply. The permanent magnet generator is in a long-term operating state.

The F – 22 aircraft uses a digital flight control system. Like the B777, it also has a 3 –

degree flight control power supply and every power supply is powered by 2 dedicated permanent magnet generators placed in the brushless DC generator. F - 22 has two 65 kW brushless DC generators, and each motor has four permanent magnet generators, one of which is the auxiliary exciter of the brushless DC generator, and the other three are permanent magnet generators for flight control. The three flight control power supplies themselves have dedicated batteries, but the batteries are only used in emergencies, the power supply is powered by permanent magnet generators during normal operation. The importance and reliability of permanent magnet generators are quite high.

In fact, the short-circuit current of the above-mentioned three aircraft-specific permanent magnet generators is only about 1.3 times the rated current, so even if the motor is short-circuited, the short-circuit current will not overheat the bleed air winding, avoiding the expansion of short-circuit faults.

The key to the permanent magnet generator's ability to greatly reduce the short-circuit current is to adopt a high-reactance design, which is achieved by increasing the number of turns of the armature winding and improving the armature slot shape because the armature reaction reactance is proportional to the square of the number of series turns of the armature winding per phase, and the leakage reactance of the armature winding is proportional to the leakage permeability of the slot, so the greater the leakage of the slot, the larger the leakage reactance will be. Therefore, the characteristics of the permanent magnet generator can fully meet the fail-safe requirements.

5.3.3 Prospect of Permanent Magnet Motor in MEA Application

The development of the dedicated permanent magnet generator and its successful application in the aircraft provide the feasibility of the application of the main generator in the aircraft power supply. Therefore, scholars at home and abroad have carried out a lot of research on the application of the permanent magnet motor in the main power supply of the MEA.

Although high reactance is still the primary condition for the application of permanent magnet generators, in order to further improve the reliability of the main generator, the concept of developing permanent magnet fault-tolerant generators is on the agenda.

The three-phase motor should not constitute a fault-tolerant motor because if one phase is failed, the remaining two-phase motor will be asymmetrical, and the output power will be reduced. The fault-tolerant motor uses a motor with a number of phases greater than 3. Obviously, the more phases there are, the smaller the impact on the work after a phase failure. However, too many phases will complicate the system, so usually the fault-tolerant motor is four-phase or five - phase motor.

The University of Sheffield has developed a 250 kW five-phase permanent magnet fault-tolerant generator driven by the fan rotor of the aviation generator. The main technical parameters are listed in Tab. 5.3.1.

Tab. 5. 3. 1 Main technical parameters of five-phase permanent magnet fault-tolerant generator

SN	Parameter	Value	SN	Parameter	Value
1	Rated capacity/kW	250(1 050—3 100 r/min)	7	Armature elements number	20
2	Armature winding	Concentrated winding	8	Permanent magnet poles number	28
3	Motor air gap/mm	3	9	Number of series elements per phase	4
4	Cooling method	Gas cooling	10	Emergency capacity/kW	25
5	Number of stator teeth	40	11	Operating temperature/℃	140
6	Phase number	5	12	—	—

It should be noted that the fan rotor of the turbofan engine is also called the low-pressure rotor. The operating speed is 1 050—3 100 r/min, and the operating speed range is $n_{max} : n_{min} \approx 3 : 1$. Another feature of the fan rotor is that when the engine fails, as long as the aircraft is still flying, the fan is still rotating with the help of the oncoming airflow. At this time, the motor can be used as an emergency generator, so the RAT emergency generator can be omitted. However, when the speed of the fan has dropped to about 250 r/min, the power of the motor drops to about 25 kW.

5. 3. 4 Development of Rare Earth Permanent Magnet Motor

Rare earth permanent magnet motors are widely used due to the excellent properties of rare earth permanent magnet materials. Rare earth permanent magnet motors have the advantages of small size, light weight, high power density, high efficiency and strong overload capacity.

The application of rare earth permanent magnet generators in aircraft has gradually developed from small power to large power and reached high power. With the deepening of people's understanding of rare earth permanent magnet motors and the birth of high reactance permanent magnet generators, not only the auxiliary exciter of the three-stage brushless alternator but also dedicated power generator for flight control and generator control system use rare earth permanent magnet motors because of the reliability of rare earth permanent magnet motor. The birth of the rare earth permanent magnet fault-tolerant generator will undoubtedly develop the permanent magnet generator to a new stage.

Rare earth permanent magnet generators are more widely used in aviation. BLDCMs, drive motors for oil pumps and fans, servo motors, high-speed motors and ultra-high-speed motors all have rare earth permanent magnet motors. In these applications, the servo motor is more important, and the EMA and the EHA are inseparable from the servo motor. The birth of the Halbach motor has created conditions for the development of high-performance servo motors, and also contributed to the development of high-efficiency and low-loss motors, as well as the development of fault-tolerant motors.

5.3.5　Dedicated Power Converter for Flight Control and Engine Control

The flight control and engine control systems are directly related to the safety of the aircraft, and the power supply of the control system is directly related to its operating capability. Therefore, the dedicated power supplies for flight control and engine control are fault-tolerant power supplies with uninterrupted redundancy. There are three sources of electrical energy for the dedicated power supply: the first one is dedicated permanent magnet generator, which is directly driven by the engine; the second source is from the aircraft grid, which is a 28 V DC grid; the third source is a dedicated battery, the battery can be in charging mode or fully charged with no-load mode when the power supply is normal, and converted to discharge mode when other power sources fail.

The power converter in the dedicated power supply is used to integrate the input of each power supply and output high-quality power. There are two types of converters: non-isolated converters and isolated converters. The non-isolated converter is used to convert the electrical energy of the dedicated generator and the storage battery; the isolated converter is used to convert the electrical energy of the aircraft grid.

The permanent magnet dedicated generator is driven by the engine, and its rotational speed has a wide variation range, and the variation range of the generator voltage is even wider. The converter of the dedicated generator is divided into two types: AC/DC and DC/DC. The former converts the alternating current of the generator into direct current, and the latter converts the changing direct current into a stable direct current. There are multiple dedicated generators on the aircraft. In general, the dedicated generators offer power to the flight control system together with the launching and controlling computer. AC/DC converters mostly use diode rectifier bridges, when DC/DC converters use step-down Buck converters.

The dedicated battery rear converter is a bidirectional DC/DC converter, which can charge the battery and convert the battery power into the power required by the control system. The converter contains a battery state monitoring circuit for converter control. In general, the battery should be fully charged and then at no load state.

The electrical energy obtained from the aircraft power grid is used to supply power to the control system when the engine is not operating and the generator has no output. It should be an isolated DC/DC converter and the power conversion is unidirectional. That is, it only allows the grid power to be converted into power of the control system.

The special generator converter, the special battery converter and the grid power converter constitute the redundant uninterrupted power supply. Each power supply is output through a reverse current protection diode to prevent a failure of power supply harming the entire power supply. High power density, high efficiency, high reliability and modular structure are the basic requirements for converters.

5.4 Summary

① The application of electric actuating mechanism creates conditions for further optimization and improvement of the aircraft, thereby realizing the replacement of the centralized hydraulic energy system. In fact, before the A380, the airbus has begun to use electric actuating mechanisms on other types of aircraft. The electric actuating mechanism does not need long hydraulic pipelines and various valves, which prevents the hazard of hydraulic oil leakage.

Electric actuation only works and gets electrical energy when it needs to be actuated. Therefore, there is no waste of electrical energy.

② The electric pump and electric fan motor in the variable frequency AC system are asynchronous motors. When the asynchronous motor is powered by a constant voltage variable frequency power supply, it is impossible to have high efficiency within the frequency range of the power supply, except when the fuel pump is at high speed. In addition to requiring high motor speed and supplying more fuel, electric fans do not necessarily require a large amount of ventilation at high engine speeds. At the same time, pumps and fans driven by asynchronous motors directly powered by the grid cannot be intelligently controlled.

③ Rare Earth Permanent Magnet Motor is widely used due to the excellent properties of rare earth permanent magnet materials. The rare earth permanent magnet motor has the advantages of small size, light weight, high power density, high efficiency and overload capacity. The application of permanent magnet motors in aircraft has gradually developed from low power to high power. BLDCM, drive motors for oil pumps and fans, servo motors, high-speed and ultra-high-speed motors all widely use rare earth permanent magnet motors. In these applications, the servo motor is more important, and the EMA mechanism and the EHA mechanism are inseparable from the servo motor.

5.5 Exercises

1. What are the applications of electric motors in aircraft?
2. What are the advantages of operating with electrical energy?
3. What are the typical applications of electric power manipulation technology in aircraft flight control and manipulation technology?
4. What are the two operating modes of EBHA?
5. What are the characteristics of rare earth permanent magnet materials?

Chapter 6　Environmental Control System of MEA

　　With innovations in MEA technology, modern large airliners are becoming safer, more economical, more comfortable and more environmentally friendly. Civil aircraft will have higher and higher requirements for operation, mainly in areas of cost, technical performance, comfort, and exhaust emissions. Since it is necessary to maintain a comfortable environment, the right temperature, fresh air, and the right atmospheric pressure for the crew and passengers throughout the flight, and due to the rapidity of the aircraft, the climate conditions change rapidly, and the adjustment of the environmental conditions must also be fast.

6.1　High Altitude Environment

　　The Environmental Control System(ECS) must deal with a wide temperature range, proper humidity, and sufficient concentration of oxygen to let the occupants on board breathe air. Civil aircraft carry a large number of passengers, and the people carried cannot tolerate extremely uncomfortable conditions under any circumstances. Appropriate ambient temperatures are also required for avionics, fuel and hydraulic systems, and the environmental control system also needs to provide functions such as defogging, anti-icing, anti-overload and raindrop removal. Therefore, the atmospheric conditions of the aircraft's work will be introduced firstly.

6.1.1　Atmospheric Physical Properties

　　Atmospheric physical properties refer to the law that the pressure, density, and temperature of the atmosphere change with altitude. The standard atmosphere is a "model atmosphere" promulgated by an authoritative organization. According to the measured data, the average vertical distribution of parameters, such as atmospheric temperature, pressure and density, is approximately expressed in a simplified way. The international standard atmosphere is promulgated by the international organization and the national standard atmosphere is promulgated by the country. The standard atmosphere is used as the basis for calibrating aircraft navigation instruments and comparing aircraft performance.

　　The general international standard atmosphere stipulates that the atmosphere is still, and the air is an ideal gas that is dry and clean. The main constants of the physical properties of the sea level atmosphere are listed in Tab. 6. 1. 1.

Tab. 6. 1. 1 Physical properties of the atmosphere at sea level

Properties	Value	Properties	Value
Temperature t_0/℃	15	Air density m_0/(kg · m^{-3})	1. 225
Air pressure p_0/mmHg	760	Speed of sound a_0/(m · s^{-1})	340. 294
Standard gravitational acceleration g_0/(m · s^{-1})	9. 806 65	Gas constant of dry air R/[J · (K · kg)$^{-1}$]	287. 053

Atmospheric pressure, density, and temperature show a downward trend with the increase of altitude. Fig. 6. 1. 1 shows the relationship between ambient temperature and altitude, Fig. 6. 1. 2 shows the curve of ambient pressure with altitude, and Fig. 6. 1. 3 shows the curve of air density ratio with height.

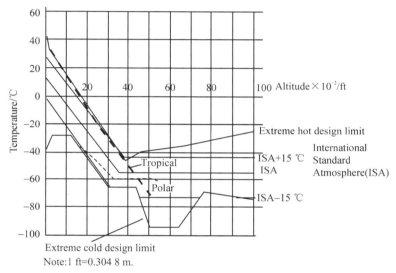

Fig. 6. 1. 1　Relationship between ambient temperature and altitude

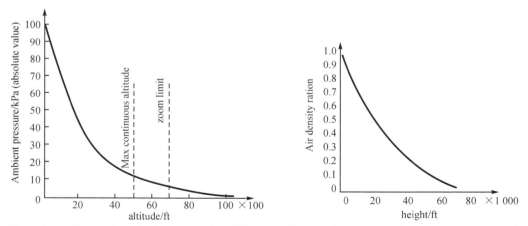

Fig. 6. 1. 2　Curve of ambient pressure with altitude　　**Fig. 6. 1. 3　Curve of air density ratio with altitude**

As listed in Tab. 6. 1. 2, it is the relationship data between temperature, pressure, density and altitude, it will enter the stratosphere when it reaches the sky above 10 000 meters, and the temperature is about −56 ℃.

Tab. 6.1.2　Relationship between temperature, pressure, density and altitude

Altitude A/m	Temperature		Pressure			Density	
	T/K	$t/℃$	p/kPa	$p/mmHg$	p/p_0	$m/(kg \cdot m^{-3})$	m/m_0
−1 000	294.65	21.50	113.93	854.55	1.124 4	1.347 0	1.099 6
−500	291.40	18.25	107.47	806.15	1.060 7	1.284 9	1.048 9
0	288.15	15.00	101.325	760.00	1.000 0	1.284 9	1.000 0
1 000	281.65	8.50	89.876	674.12	0.887 0	1.111 7	0.907 5
2 000	275.15	2.00	79.501	596.30	0.784 6	1.006 6	0.821 7
3 000	268.66	−4.49	70.121	525.95	0.692 0	0.909 3	0.742 3
4 000	262.17	−10.98	61.660	462.49	0.608 5	0.819 4	0.668 9
5 000	255.28	−17.47	54.048	405.39	0.533 4	0.736 4	0.601 2
6 000	249.19	−23.96	47.217	354.16	0.466 0	0.660 1	0.538 9
7 000	242.70	−30.45	41.105	308.31	0.405 7	0.590 0	0.481 7
8 000	236.22	−36.93	35.651	267.40	0.351 9	0.525 8	0.429 2
9 000	229.73	−43.42	30.800	231.02	0.304 0	0.467 1	0.381 3
10 000	223.25	−49.90	26.499	198.76	0.261 5	0.413 5	0.337 6
11 000	216.77	−56.38	22.699	170.26	0.224 0	0.364 8	0.297 8
12 000	216.65	−56.50	19.399	145.50	0.191 5	0.311 9	0.254 6
13 000	216.65	−56.50	16.579	124.35	0.163 6	0.266 6	0.217 6
14 000	216.65	−56.50	14.170	106.28	0.140 0	0.227 9	0.186 0
15 000	216.65	−56.50	12.111	90.85	0.119 5	0.194 8	0.159 0

Note: 1 mmHg=133.322 368 4 Pa.

6.1.2　Effects of High-Altitude Environment on Human Physiology

1. High Altitude Hypoxia

As the flight altitude increases, the atmospheric pressure decreases, the partial pressure of oxygen in the atmosphere and the partial pressure of oxygen in the alveolar air will also decrease accordingly. Then the oxygen saturation in the blood will decrease, and the body tissue cells cannot get a normal oxygen supply. There are various discomforts in the human body: headache, sluggish reflection, poor hearing, vision loss, emotional disturbance, purple lips and nails, etc.

2. The Danger of Low Pressure

The physiological effects of atmospheric pressure changes on the human body are mainly reflected in the hazards of hypoxia, the symptoms of swelling and emphysema caused by low pressure, and the hazards of explosive decompression caused by rapid pressure changes. With the reduction of atmospheric pressure, the human body will appear high altitude flatulence and decompression sickness.

3. Rate of Pressure Change and Hazards of Explosive Decompression

If the pressure change rate is too large, tinnitus, dizziness, and nausea will occur. The human body is more sensitive to the excessive pressure increase rate, so when the plane descends, the earache is more serious. Explosive decompression is the phenomenon when the cockpit is suddenly lost at high altitude, the pressure change rate is extremely large, which will cause great harm to the human body. The safety measures after the explosive decompression accident occurred are as follows: the aircraft was quickly lowered to a safe altitude of about 4 000 m, and the oxygen equipment was used as soon as possible.

For example, when the altitude increases to 19. 2 km, the atmospheric pressure drops to 47 mmHg. Under this pressure, the boiling point of water is 37 ℃, which is the body temperature. Under this pressure, the body fluids boil, causing tissue swelling and damage to the human body. This phenomenon is called high altitude decompression sickness.

(1) Influence of Temperature and Humidity

Ambient temperature and humidity have a great influence on the balance of temperature and moisture in the human body. The suitable temperature for the human body is 15— 25 ℃. The main effect of humidity on the human body is dryness, and beverages need to be supplied.

(2) Other Influencing Factors

The cleanliness of the cabin air means that it is free of impurities and harmful gases and the ventilation of the sealed cabin. The jet noise from the engine and the ventilation noise are all factors that affect the comfort of the human body.

In order to overcome the hazards of the air environment, feeding devices and airtight cockpits are often used, so that the pressure in the cabin is greater than the atmospheric pressure outside, and the environmental parameters of the cabin are adjusted to create a comfortable cabin environment and meet the physiological and work needs of the human body.

6.2 Electric Environmental Control System

An electric Environmental Control System (ECS) is used to maintain temperature, pressure and humidity in the cabin and cockpit, reduce particulate matter in the air, keep the air fresh, and provide cooling for high-power equipment and avionics and electrical equipment in the electronic and electrical equipment bay.

The environmental control system consists of air conditioning system, air cooling system, cabin air recirculate system, temperature regulation and control, humidity control, ventilation control, and avionics and electrical equipment cooling system.

6.2.1　Air Conditioning System

1. The Role of the Air Conditioning System

The cockpit air conditioning system provides a good environment for the cockpit, passenger cabin, equipment cabin and cargo hold of the aircraft under various flight states and external conditions, so as to ensure the normal operating conditions and living environment of the pilots and passengers, the normal operation of the equipment and the safety of cargo. After the aircraft cabin is sealed, the air supply is pressurized, and the pressure inside the cabin is greater than the atmospheric pressure outside the cabin, creating a comfortable cabin environment to meet the needs of human physiology and work.

The environmental parameters of the airtight cabin are temperature, pressure, pressure change rate and ventilation rate. The air conditioning system is designed to let air enter the cabin through overhead ducts and exits through the floor. It reduces the risk of fore-and-aft cabin air flow and cross-contamination between passengers.

2. Air Distribution

The function of the air source system of a traditional aircraft is to provide a pressurized air source to the cockpit, control the pressure, flow and temperature of the supplied air, and then adjust its temperature and pressure through the air conditioning component and then supply it into the cockpit; Anti-icing heating of the engine or leading edge of the wing, etc, and pressurisation of water system, hydraulic system, etc. The types of air sources include engine compressor bleed air, supercharger bleed air, APU bleed air and ground air source vehicles. The source of engine bleed air is drawn from the compressor of the turbofan engine. There are low-pressure stage bleed air and high-pressure stage bleed air at the bleed air part. In order to reduce the loss of engine power, modern airliners use two-stage bleed air. When the low-pressure stage bleed air is insufficient, it can be supplemented with the high-pressure stage bleed air. At this time, the low-pressure stage has Non Return Valve (NRV) to prevent back flow.

The air in the unconditioned bays is not completely still, and the aircraft aisles are designed to have a constant airflow through each equipment bay. Especially in rapid climbs and descents, no pressure difference is established between the cabins, and the ventilation air flow is often the outlet air flow of the conditioning cabin. Cabin distribution systems are designed to provide the most comfortable environment possible without hot spots, cold spots or wind. Since passengers can move around freely, there must be good comfort throughout the cabin and they can control their own vents and flow and airflow direction. Modern large aircraft provide general air conditioning and no longer provide individual vents due to the high performance of the air conditioning system.

The cross-transmission air supply is controlled by the cross-transmission valve, and any engine bleed air can be used for any air-conditioning system to work and the engine to start.

The main disadvantage of alternate air supply is contamination of the air supply may influence engine performance (power loss). It is therefore necessary to develop electric air distribution, which no longer bleeds air from the engine.

3. Electric Environmental Control System of MEA

Fig. 6. 2. 1 shows the typical cabin division of the B787 cockpit temperature control system. The cabin is divided into zones 1 to 4, and the temperature can be set separately in the cabin. There are two air conditioning components in the middle of the B787 fuselage. There are two electric Cabin Air Compressors(CAC) in each air conditioning unit. The CAC inside the aircraft is called CAC1, and the outside one is called CAC2.

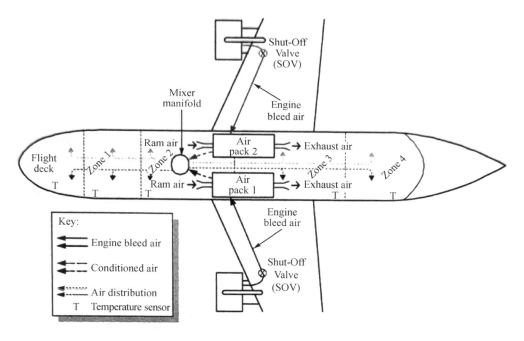

Fig. 6. 2. 1 Cabin temperature control system (B787)

The MEA adopts an electric ECS, which cancels the bleed air from the engine. It consists of a closed electric air circulation system, an electric closed evaporate circulation system, and a liquid cooling system. Its core is an electric turbine air circulation refrigeration system.

The electric turbo refrigeration system is a refrigeration system with an electric turbo cooler as the core accessory. Compared with the traditional turbo refrigeration system, it has the following two characteristics:

① The air under normal pressure can be expanded to do work after being pressurized and reduce the temperature and increase the application range of the turbo refrigeration system.

② The purpose of adjusting the turbo refrigeration power can be achieved by controlling the motor, so that the refrigeration system can always work in an optimal state.

As shown in Fig. 6.2.2, it is the diagram of the air-conditioning assembly (B787). Each air conditioning assembly is composed of two electrically driven compressors, one ozone converter, two air/air heat exchanges, a set of air circulation and mixer, etc.

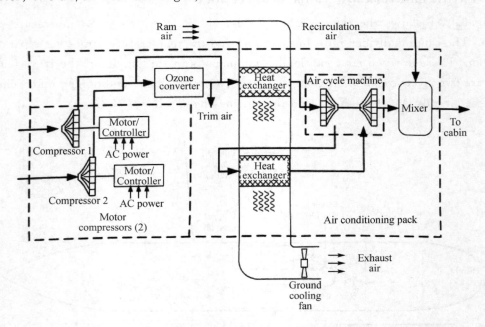

Fig. 6.2.2 Air-conditioning assembly of electric environment control system (B787)

The electric power required by the more-electric ECS is very high. Each electric compressor adopts a 125 kV · A permanent magnet motor, which is controlled by the motor controller located in the electronic and electrical equipment compartment. The main benefit is that the air is no longer extracted from the central duct of the engine, and more importantly, the temperature and pressure of the delivered air are much lower.

As shown in Fig. 6.2.3, it is a comparison diagram of bleed air and electric drive environmental control system. It can be seen from the figure that in the conventional environmental control system of bleed air supply, the typical temperature of bleed air entering the engine is 204.4 ℃ (400 ℃), pressure is 30 psi(1 psi = 6 894.757 Pa). After cooling by the air circulation, the typical output is about 15.6 ℃ (60 ℉) and 11.8 psi, which is equivalent to atmospheric pressure at a cabin altitude of about 6 000 feet (1 524 m).

The electric drive compressor raises the air pressure to a lower 15 psi, the temperature is 93.3 ℃. After the compressed air is cooled by the No.1 air/air exchange, it enters the compressor of the air circulation machine. After being cooled by the No.2 air/air heat exchange, the air at the outlet of the compressor is cooled again by the expansion turbine, and the cooled air is mixed with recirculated air from the cabin, the temperature of the mixed air is 11.8 psi, the pressure is 11.8 psi, and the corresponding ambient temperature is controlled at 15.6 ℃, which makes the occupants more comfortable.

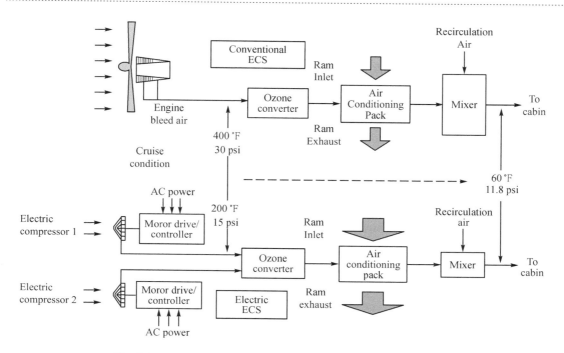

Fig. 6. 2. 3   Comparison diagram of bleed air and electric drive ECS

The two compressors in Fig. 6. 2. 3 convert electrical energy into air pressure energy. The air pressure and flow at the motor output are the loads of the motor. The change law in the amount of energy obtained by the electric compressor from the power grid and other electrical parameters under different flow rates should be studied. It is necessary to study the influence of electrical loads on the power supply from the perspective of utilizing electrical energy.

As shown in Fig. 6. 2. 4, the electric Cabin Air Compressor(CAC) is composed of high-speed and high-power electric motor, centrifugal compressor, adjustable diffusion actuator, compressed air inlet and outlet, etc. There will also be temperature, pressure and flow sensors and a Pack Controller Unit(PCU) in the system for data collection and control. The electric CAC is the heart of the cabin ventilation system, providing hot and high pressure air for cooling components and ambient clean air systems. This high pressure air passes through the ozone converter and cooling assembly as input air to drive the air circulation machine.

The high-speed high-power electric motor in the CAC, supported by air bearings, operates at very high speeds. Adjustable diffusion actuators ensure efficient operation of compressor and prevent surges.

The electric power of the MEA compressor is large, and its power supply is provided by the variable frequency voltage through Common Motor Start Controller(CMSC). The power supply of the CMSC is 540 V HVDC, which comes from ATRU on the aircraft. The control commands of the CMSC come from the air-conditioning Pack Controller Unit(PCU), which is used to change the motor speed to adjust the outlet pressure, temperature, and flow of the

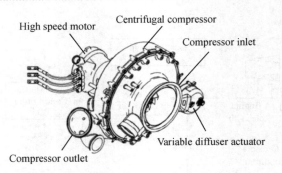

Fig. 6.2.4　Electric cockpit compressor

electric compressor. The maximum speed of the motor can reach 43 929 r/min, and its type is an air suspension motor, which is air-cooled.

The inlet damper is driven by a BLDCM. Usually, the operating voltage of the motor is 28 V DC. The inlet damper is closed on the ground. When the engine starts and reaches idle speed, the damper is opened to half of the opening. When the engine power reaches the take-off power, that is, when the speed of the high-pressure rotor reaches 90% of the maximum speed, the throttle is fully opened. When flying in the air, the damper is open even if the electric compressor is not operating. The actuation of the damper is controlled by the PCU.

Surge may occur when the compressor is operating, and the surge is related to the air flow, compression ratio, motor speed and the position of the diffuser moving vanes. Controlling the motor speed and the position of the diffuser vanes can change the compression ratio and air flow and suppress surge.

The larger the opening of the Air Heating Valve (AHV), the more air flow from the compressor outlet back to its inlet, and the higher the outlet air temperature, so AHV is the outlet temperature adjustment valve of CAC. AHV can also be used to suppress compressor surge after diffuser control failure.

The air conditioning PCU regulates the motor speed based on the inlet and outlet air temperature and flow signals. The PCU accepts the air flow sensor signal of the Cabin Electrical Air Compressor (CEAC), and is used to control the speed of the motor compressor, the opening of the inlet damper, the position of the diffuser guide vane and the opening degree of the AHV to allow CEAC to operate properly.

During normal operation, the maximum temperature of the outlet air of the Cockpit Electric Air Compressor (CEAC) is 193 ℃. If the outlet temperature of the CEAC exceeds 204 ℃, the outlet air flow of the CEAC must be reduced. If the outlet air temperature of the CEAC exceeds 218 ℃, the CEAC should be closed immediately.

6.2.2　Air Cooling System

The air cooling system is used to cool the high temperature and high pressure air from the CAC and remove moisture, and each air conditioning component has an air cooling system. The air cooling system consists of two parts: an air/air heat exchange and an Air

Cycle Machine (ACM). The cooling air for the heat exchange comes from ram air introduced through the rear opening of the aircraft fuselage. The ram air inlet damper of the air cooling system of the B787 environmental control system adopts an S-shaped elbow, and the inlet of the duct has an adjustable ram air inlet damper. After the air enters the duct, it is divided into two paths, one of which goes to the nitrogen generator at the right air conditioning unit. Nitrogen Generation System (NGS), another way into the heat exchange, and then exhausted to the outside of the machine through the outlet damper.

1. Air Cycle Machine

The ACM consists of a compressor, two expansion turbines T1 and T2 and a condenser. After the high-temperature and high-pressure air sent by the CAC enters the first-stage heat exchange for cooling, it is pressurized by the compressor of the ACM. If the Air Bypass Valve (ABV) of the ACM is closed at this time, the pressure and temperature of the supercharged air have increased again, so it should enter the second-stage heat exchange for cooling. If the Economic Cooling Valve (ECV) is closed at this time, the air will enter the condenser for cooling. The air from the condenser enters the moisture extractor, which dries the humid air, and the discharged water enters the front of the ram air line heat exchange and is ejected from the nozzle. If the Low Pressure Limit Valve (LPLV) is closed, the dry air enters the first stage expansion turbine T1, the air temperature and pressure of the T1 outlet decrease significantly, and the part of the air inlet condenser is used to reduce the temperature of the air entering the human water extraction machine. The air from the condenser outlet enters the second-stage turbine T2, and the air temperature and pressure at the T2 outlet are further reduced. T2 outlet air and cockpit back re-circulation air in the mixer mix the cockpit.

2. B787 Air Cooling System

The air cooling system of the B787 is controlled by the Park Control Unit(PCU). The sensor signals sent to the PCU cooling system are:

① Two air conditioning pack unit outlet temperature sensors.

② Two air temperature sensors sent to the cockpit.

③ Two condenser inlet and outlet temperature sensors.

④ Two second-stage heat exchange outlet temperature sensors.

⑤ ACM compressor outlet temperature sensor.

⑥ Speed signal of ACM.

The PCU controls the cooling system components, mainly including the inlet valve and outlet valve of the ram air pipeline, the ABV of the ACM, the economical cooling valve ECV and the LPLV, and the position signals of the valves are fed back to the PCU.

When the aircraft is flying, the ram air fan control valve is opened under the pressure of the oncoming air flow. Even if the electric fan does not work at this time, there is enough air flow. In order to prevent the damage of the air bearing, the electric fan is in a low-speed operating state.

The outlet damper of the ram air duct is used to control the flow of the ram air, thereby controlling the outlet temperature of the PCU and the temperature of the coolant of the Power Electric Cooling System (PECS). If the ABV and the LPLV in the ACM cannot make the air conditioning PCU work in the required range, the temperature can be controlled by controlling the position of the outlet damper. If the liquid temperature of PECS is too high, the outlet damper will be preferentially controlled to reduce its temperature.

If the aircraft is on the ground, the large inlet damper must be opened when the large outlet damper is open. Usually, the air outlet is in the fully open position, that is, at an angle of 40°. During the flight, the opening angle of the outlet damper is reduced to 20°. If both the ram air inlet damper and the outlet damper fail but are not all closed, the air conditioning PCU can control the outlet air temperature of the air conditioning pack by controlling the valve of the ACM.

The ACM further cools and dries the high temperature air sent by the CAC. ACM consists of 3 components, the No. 1 turbine is used to provide cold air to the No. 2 passage of the condenser, the compressor sends the high-temperature air at its outlet to the No. 2 heat-exchange for cooling; the No. 2 turbine sends the cold air from the outlet to the air mixer. It can be seen that the ACM and the heat-exchange work together to cool and dry the air from the CAC. The ACM has a maximum operating speed of 23 000 r/min and weighs 39 kg.

3. Air/Air Heat-Exchange

The inlet of the ram air line has an inlet damper and its actuating mechanism, as well as an inlet air heater, and the water nozzle is in front of the heat exchange. The heat exchange assembly consists of 4 parts. The first and second stages are both air/air heat exchanges. These two heat exchanges are used to cool the high temperature air sent by the CAC; the third part is used for air/liquid heat exchange for high power electronics cooling; the fourth part is the heat exchange for the air conditioning of the front cargo compartment, the heat exchange is a replaceable part and weighs 113 kg. The inlet damper of the ram air is used to control the flow of the ram air, the temperature of the PECS coolant in the cooling system of the power electronic equipment and the outlet air temperature of the air conditioning pack. If the temperature of the PECS coolant is too high, the damper should be more opened to allow cooling the temperature of the liquid was lowered to 30 ℃.

4. Fan Motor

Another part of the heat exchange is the fan motor, whose controller is not in the air duct, and below the fan is the ram air fan control valve. The valve consists of 4 flaps and 2 hinges. The ram air outlet assembly is a carbon fiber composite component that has the electric fan, inspection valve, and ram air outlet valve.

When the aircraft is on the ground, an electric fan in the ram air duct is used to circulate the air in the duct, and the air conditioning pack can work to cool the power electronic components. The electric fan motor adopts air bearing, its motor rotor and fan are coaxial, adopting air cooling. Fan motor controller or Common Motor Starter Controller (CMSC) drive the fan motor, the speed is 11 500 r/min when operating on the ground, and the speed when operating in the air is reduced to 6 000 r/min.

When the CMSC has no power supply or needs to use the CMSC to start the aero-engine, the controller of the fan motor itself is used to drive the fan motor. On the ground, if the ambient temperature is lower than −1 ℃, or the coolant temperature of the PECS is higher than 68 ℃, when the aircraft is in the air, the fan motor controller is used to drive the motor. The fan motor is controlled by the PCU and cooled by the PECS system.

5. Air Circulation Machine Air Flow Valve Control

The air cycle system of the B787 is composed of two expansion turbines, a condenser and a compressor. The air conditioning PCU controls the air flow of the ACM by controlling the three valves of LPLV, ECV and ABV.

(1) Low Pressure Limit Valve Control

In order to prevent water from freezing in the condenser, the LPLV is used to control the temperature of the air in the condenser passage, the cold air from the turbine and the hot air passing through the LPLV are mixed in the pipeline, making the inlet temperature of the condenser not lower than 0 ℃.

If the air is relatively dry, even if the temperature is low, it is not necessary to prevent the water from freezing in the condenser to close the LPLV, so the PCU uses the detected data to control the LPLV's actuating mechanism after calculation.

(2) Economical Cooling Valve

Open the ECV to bypass the condenser, so that the air sent by the cabin air compressor CAC directly enters the turbine, so as to reduce the power consumption of the air conditioning system.

(3) Air Bypass Valve

The condenser bypass operating mode can also be entered under abnormal operating conditions, that is, when the ACM fails or the ACM compressor inlet temperature sensor fails.

In the condenser bypass state, the PCU controls the outlet air temperature of the air conditioning assembly by controlling the stamping air population valve, the outlet valve and ACM, and the ABV.

Normally, the compressor outlet temperature is 149 ℃, and the PCU adjusts the air flow to control the compressor outlet temperature. If the compressor outlet temperature exceeds 218 ℃, the PCU will turn off the air conditioner after 10 s. The air conditioning unit outlet temperature limit is 82 ℃. If the temperature is too high, the PCU will turn off the air conditioning pack.

6.2.3　Cabin Air Re-circulation System

The cabin air re-circulation system includes re-circulation subsystems, air filters, re-circulation fans, and air cooling devices.

1. Re-circulation Subsystem

The cabin air re-circulation system passes the humid air in the cabin through the dryer at the top of the cabin and then enters the mixing unit. Air from the front electrical and electronic equipment compartment also enters the mixing unit, then passes through the air filter, High Flow Shut-Off Valve (HFSOV), re-circulation fans, and air/liquid heat exchanges and enters the air mixer. The air is mixed with air from the air conditioning unit and then sent to the cabin air distribution system. The air re-circulation system is used for ventilation and reuse of cabin air, reducing the load on the air conditioning components and maintaining reasonable cleanliness and humidity.

2. Air filter

The air filter is used to filter particles and infectious substances from the air. It is important to note that the filter should be cleaned regularly.

3. Re-circulation Fan

A re-circulation fan is used to extract air from the cabin into the air re-circulation system. The power supply of the electric fan is 235 V variable frequency AC power supply, the motor is an asynchronous motor with a simple structure, the fan and the motor are coaxial, and the bearing uses an air bearing. The total weight of the fan is 21 kg.

There is a fan switch on the P5 board of the cockpit head panel, and the fan works when the switch is closed. The fan stops operating when the fan fails or another circulation system works or a fire alarm occurs. When the aircraft is on the ground and the external power supply is turned on, if the two air-conditioning components are not operating, even if the switch is in the off position, the fan will still work, and the fan controller is used to adjust the motor speed to adjust the air flow.

4. Air Cooling Device

The outlet air from the fan enters the air/liquid heat exchange for cooling, and the cooled air enters the mixer and is mixed with the air from the air conditioning unit. The coolant comes from the Integrated Cooling System (ICS), a non-toxic mixture of acrylic glycol and water in a 60:40 ratio. The coolant enters the heat exchange when the switching valve of the ICS is opened.

The mixer has two air inlets: one is the recirculated air from the heat exchange and the other is the outlet air from the air conditioning pack. The mixer has two air outlets: one for the air distribution system to the cockpit, and the other for the air distribution system to the cockpit. When the air conditioning pack is turned off, the inlet air to the mixer is fed with recirculated air.

6.2.4 Temperature and Humidity Regulation and Control

Temperature regulation and control mainly includes regulation and control of cockpit temperature and over-temperature detection and protection of the air-conditioning pack.

1. Cockpit Temperature Adjustment and Control

As shown in Fig. 6.2.5, it is the cockpit heating temperature regulator panel, there is an air conditioning system on the control screen of the overhead panel P5 in the cockpit, and there are cockpit temperature and cockpit temperature adjustment knobs on the ECS control surface. There is a special heater temperature adjustment panel in the cockpit, as well as a special control panel for cabin temperature adjustment, etc.

Cabin temperature adjustment is an important part of the environmental control system, which is adjusted according to the cabin temperature set value and the detection signal of the temperature sensor in the cabin. Fig. 6.2.6 shows the distribution of cockpit temperature sensors.

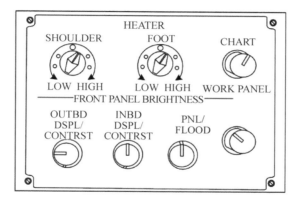

Fig. 6.2.5 Cockpit heating temperature regulator panel

There are various sensors in the cargo compartment heating subsystem and the pressure control system, among which there are many temperature sensors. As listed in Tab. 6.2.1, it is the temperature sensor in the B787 environmental control system.

Tab. 6.2.1 Temperature sensors in B787 environmental control system

SN	Detection position(T)	Number	Interface	SN	Detection position(T)	Number	Interface
1	CAC compressor outlet	2	PCU	5	Air conditioning components export	2	PCU
2	No.. 2 heat exchange outlet	2	PCU	6	A1 area pipeline	1	RDC15
3	Condenser inlet	2	PCU	7	A2 – 2 area pipeline	1	RDC15
4	Cockpit air supply	2	PCU	8	A2 – 1 area pipeline	1	RDC16

(Continued)

SN	Detection position(T)	Number	Interface	SN	Detection position(T)	Number	Interface
9	B2 – 1 area pipeline	1	RDC16	18	Cockpit A1 area	1	RDC16
10	B2 – 2 area pipeline	1	RDC21	19	Cockpit A2 area	1	RDC14
11	B1 area pipeline	1	RDC21	20	Cockpit B1 area	1	RDC21
12	C-area rear pipeline	1	RDC16	21	Cockpit B2 area	1	RDC13
13	C-area rear pipeline	1	RDC9	22	Cockpit area C	1	RDC10
14	Cabin B1 rear line heater	1	RDC21	23	Cockpit area D	1	RDC9
15	D -area front pipeline	1	RDC9	24	A1 cabin line heater	1	RDC13
16	D -area rear pipeline	1	RDC21	25	Ceiling temperature	2	RDC10
17	Cockpit	1	RDC11,12	26	—	—	—

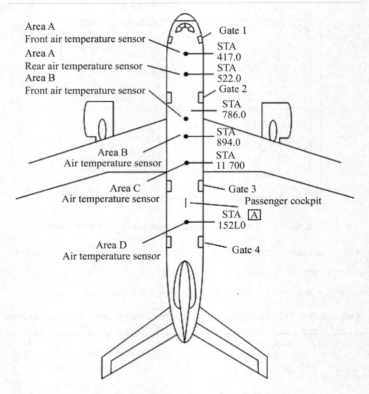

Fig. 6. 2. 6   Distribution of cockpit temperature sensors

The signal of the cockpit temperature sensor is sent to the data bus through the Remote Data Concentrate(RDC), and the operating parameters of the PCU of the air conditioning assembly are sent to the Common Core System(CCS) through the data bus. As listed in Tab. 6. 2. 2, the electric motor used in the Cabin Air Compressor (CAC) of the air-conditioning pack, the total number of actuating mechanisms of the CAC is 10.

Tab. 6. 2. 2　Actuating mechanism of CAC

SN	Name	Quantity	Motor type	Controller
1	CAC intake valve	1(Each component)	BLDCM	PCU
2	CAC diffuser	1(Each CAC)	BLDCM	PCU
3	Air Heating Valve (AHV)	1(Each HAV)	BLDCM	PCU

As listed in Tab. 6. 2. 3, it is the electric motor for air cooling system, including various actuators, control valves, regulating valves and other electric actuating mechanisms. These actuating mechanisms are controlled by the Pack Controller Unit (PCU) of the air-conditioning pack. The PCU operates according to the temperature and pressure signals detected by the sensor and the P5 command sent by the RDC through the data bus. The operating parameters of the air-conditioning pack are sent to the CCS through the data bus, and displayed by the display. The display not only displays the operating parameters, but also displays the operating status of the system.

Tab. 6. 2. 3　Electric motors for air cooling systems

SN	Name	Number	Controller	Remark
1	Ram air inlet valve actuator	1	PCU	
2	Ram air outlet valve actuator	1	PCU	
3	Ram air fan motor	1	Ram air fan motor controller	
4	LPLV actuator	1	PCU	
5	Economical Cooling Valve (ECV) actuator	1	PCU	Quantity of each air conditioning unit
6	ACM bypass valve actuator	1	PCU	
7	Cabin air re-circulation system fan	1	PCU	
8	Adjusting the air pressure regulator valve actuator	1	PCU	
9	Adjusting the air valve actuator	3	PCU	
10	Cockpit charge air valve actuator	1	PCU	
11	Cockpit charge air fan motor	1	PCU	

The temperature sensor is the source of obtaining the temperature information of the environmental control system equipment and is the foundation of information processing and judgment. For details, please refer to the relevant information.

The cabin temperature adjustment is based on the cabin temperature setting value and the detection signal of the temperature sensor in the cabin. There are cabin temperature and cockpit temperature adjustment knobs on the control box of the cockpit overhead power control panel P5 for adjustment. Tab. 6. 2. 4 lists the auxiliary heaters in the cockpit and cabin.

Tab. 6. 2. 4 Auxiliary heaters in the cockpit and cabin

SN	Name	Number	Power/W	Remark
1	Cockpit stand surface heater	2	50	
2	Cockpit upper back duct heater	2	1 000	
3	Cabin aisle duct heater	7	1 500	Each one
4	Door area floor panel heater	4	—	

2. Over-temperature Detection of Air Conditioning Pack

The temperature detection of the air-conditioning pack cabin and the Nitrogen Generation System (NGS) is very important, so an over-temperature detection and protection system is set up. Damaged pipeline seals, disengagement of pipeline welds, and cracked joints can cause high temperature gas leakage and cause overheating. Therefore, the temperature in the vicinity of aircraft structures, cables, fuel lines and hydraulic lines should be checked. The B787 has 5 dual-sensor temperature detection loops, one for the nitrogen generation system, and each air conditioning pack has 2 sensor loops, and the over-temperature signal is sent to the over-temperature protector with dual channels.

The self-checking period of the over-temperature protection subsystem is 15 minutes. When one of the dual sensors is found to be faulty, the system switches to receive the over-temperature signal of only one sensor to realize protection. There is a phase change material in the temperature sensor. When the temperature is too high, the resistance of the phase change material between the outer conductor and the inner conductor of the sensing element decreases sharply, and the resistance returns to the original state after the temperature decreases. The maximum temperature of the air conditioning pack cabin during normal operation is 71 ℃. If the temperature exceeds 99 ℃, the controller of the over-temperature detection and protection subsystem will send a signal to the CCS. CCS will shut down air conditioning components in 20 seconds.

Fig. 6. 2. 7 shows the ECS control panel of the air conditioning system of the overhead panel control panel P5 of the B787. Below the cockpit stripe knob is the left air conditioning component button, and below the cabin temperature adjustment knob is the right air conditioning component button. When the power is on, press the button to work.

As listed in Tab. 6. 2. 5, they are the three operating modes of the PCU and CCS control air conditioning pack of the B787, which are placed in the air-conditioning system control box of the P5 control screen in the cockpit of the B787.

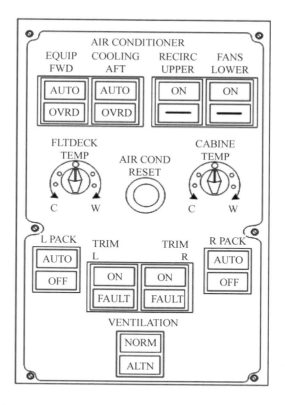

Fig. 6. 2. 7　Air conditioning system ECS control panel P5(B787)

Tab. 6. 2. 5　Three operating modes of the PCU and CCS control air conditioning pack of the B787

SN	Name	Describe
1	Heat exchange only	Used when controlling ACM failure, the ECV and ABV open the PCU to regulate the outlet air temperature of the air conditioning assembly by controlling the ram air inlet and outlet valves
2	Condenser bypass operation	Only used when the condenser input temperature sensor fails and the ECV cannot be shut down
3	Economic cooling	During the flight, the PCU turns the ECV on

　　If the air conditioning pack trips during operarion, it must be protected, as listed in Tab. 6. 2. 6, it is setting temperatures of the B787 Environmental Control System(ECS) air conditioning pack for trip protection. If the PCU fails to cause overheating, the air conditioner control box automatically sends a signal to the motor controller to shut down the CAC. When the temperature returns to below 60 ℃, pressing the "AIR COND RESET" button of the control box can restart the air conditioning pack.

Tab. 6. 2. 6   **Setting temperatures of the B787 ECS air conditioning pack for trip protection**

SN	Trip protection condition	Remark
1	Air Cycle Machine (ACM) outlet temperature is too high	—
2	ACM compressor outlet temperature is too high	—
3	Air conditioning unit outlet temperature is too high	—
4	Over-temperature detection and protection subsystem sensor outputs over-temperature signal or CMSC is faulty	Temperature limit 99 ℃
5	ACM compressor outlet temperature is the limit value	218 ℃
6	Air conditioning unit outlet temperature limit	88 ℃

3. Humidity Control

Temperature control, cabin pressurisation, and humidity control are all for passenger comfort, as the moisture in the air decreases with altitude, for example, in northern Europe, the typical air moisture content is 10 g/kg, while in some parts or some areas or the far east have moisture content of 30 g/kg. In hot and humid climates, cabin comfort controls are relatively complex, requiring a lower supply of air temperature at the cabin inlet. Without good humidity control, it can lead to the introduction of damp mist into the cabin. The electrical and electronic equipment does not allow excessive condensation, the humidity value must be controlled, and humidity control also reduces the requirements for windshield and window glass defogging and anti-fogging systems.

The B787 dehumidifying system is placed before and after the ceiling and consists of two dryers and pipelines. The dryer consists of an air filter, an electric fan, an electric heater, a drying wheel and a controller.

When the fan of the dryer works, the air in the ceiling area is sucked into the dryer through the air filter, and the air is discharged through the heater and the drying wheel. The drying wheel is made of glass fiber with a honeycomb structure filled with silica gel. The rotating speed is 2 r/min during operation, which separates the dry and wet air, and the electric heater heats the air entering the dryer.

The controller actually has the function of a power converter, which converts the 235 V, 360—800 Hz variable frequency AC main power supply into 115 V, 400 Hz AC for the motor, and provides power for fans, heaters and drying wheels through the control of the controller components. It carries out control and protection, monitors the outlet temperature of the dryer, and carries out over-temperature protection, reports the operating parameters to the CCS, and the data provided by CCS is the basis for the controller to work.

As listed in Tab. 6. 2. 7, it is the over-temperature protection limit value of the dryer of the B787 environmental control system. The drying air output of the dryer is about 4. 8 m^3/min, and the wet air flow is 1 m^3/min.

Tab. 6. 2. 7   The over-temperature protection limit value of the dryer of the B787 environmental control system

SN	Position	Temperature/℃	SN	Position	Temperature/℃
1	Outlet air	85	3	Fan motor	130
2	Drying wheel rotor	71	4	Heater	121

When the dryer outlet temperature is higher than 130 ℃, the CCS will turn off the heater. If the ceiling area temperature is higher than 49 ℃, CCS will turn off the electric heater; if the ceiling area temperature is higher than 52 ℃, CCS will turn off the fan. After the dryer is turned off, the fan and drying wheel continue to work for a while, causing its internal temperature to drop faster.

6.2.5   Ventilation Control

Ventilation control mainly includes toilet, kitchen, cockpit, crew rest cabin, passenger cabin, large cargo hold and cockpit ceiling. The exhaust ducts placed in the roof of the aircraft extend all the way to the large cargo compartments in the lower rear of the aircraft to allow the air from these compartments to be exhausted outwards. The equipments in the bulk cargo hold line are the air filter, the air/liquid heat exchange for the toilet/galley ventilation subsystem (one on each side) and 2 ventilation fans. When the fan is operating, the air in the pipeline can be exhausted to the Outflow Valve (OFV) at the rear of the aircraft. The OFV is on the aircraft fuselage and can exhaust air to the outside of the aircraft when it is turned on.

The exhaust line is a section of unsealed pipe with 3 safety nozzles, 2 safety nozzles for roof ventilation, and 1 safety nozzle for large cargo hold ventilation. These 3 safety nozzles ensure that the ventilation pipeline has sufficient exhaust air volume to meet the needs of liquid heat dissipation of the air/liquid heat exchange.

The CCS can control the speed of the fan through the motor controller installed on the fan. The fan motor uses the 230 V AC main power supply to work, and the fan motor has three operating modes which are as follows:

① In normal operating mode, only one fan motor works at this time.

② In operating mode of motor rotor acceleration, the two fans work at the same time. This is only the case when the following three conditions occur at the same time, that is, the aircraft is in the take off condition, one engine is damaged, and all the cabin air compressor CAC are not operating. It also requires two fans to work at the same time when the AVS is operating.

③ When the fire alarm in the rear cargo compartment and the fan motor fail, both fan motors will not work.

6.2.6   Cabin Pressure Control

The B787 has two front and rear pressure control systems. The pressure control system consists of three main components: the Outflow Valve (OFV), the Outflow Valve Controller (OFVC) and the Remote Sensor Unit (RSU).

1. Outflow Valve(OFV)

There are two outflow valves, namely the front outflow valve and the rear outflow valve. When it is normal, the front outflow valve flows out 30% of the air, and the rear outflow valve outflows 70% of the air. The outflow valve has two doors, the larger door is only open when the plane is on the ground, and the smaller door is used to adjust the cabin pressure during the flight. Each outflow valve is controlled by three electric mechanisms, one of which is used to automatically control the valve, one motor is used for main use, and the other is used for backup. The automatic control motor is a BLDCM, and the manual control is a brushed DC motor.

2. Outflow Valve Controller

There are two control channels in the outflow valve controller unit VCU, which are powered by different RPDUs respectively, interface with different RDCs, and receive CCS control commands and return valves' status information through the data bus. There are two pressure signal sensors in the VCU, and there are communication lines between the two VCUs. CAN bus is used to communicate between VCU and RDC.

3. Remote Sensor

The RSUs are located in the front and rear electrical equipment compartments respectively to detect the cabin pressure and send the pressure signal to CCS. The two sensors are backup for each other.

There are 2 safety valves in the cabin pressure control system of the B787 to avoid damage to the aircraft due to excessive pressure difference between the inside and outside of the aircraft. One of them is a Positive Pressure Safety Valve (PPSV) and the other is a Negative Pressure Safety Valve (NPSV). The safety valve keeps the pressure in the aircraft cabin within a safe range.

The normal pressure difference inside and outside the machine is 64.97 kPa. If the PPSV is opened when the pressure difference is 67.38 kPa, the PPSV opening signal is sent by the No. 1 sensor. If the PPSV does not open, when the pressure difference is 70.48 kPa, the No. 2 sensor sends a signal to open the PPSV. The NPSV opens when the negative pressure difference is −1.72 kPa.

6.2.7 Electrical Components of Environmental Control System

The environmental control system no longer uses engine bleed air, but uses electrical energy to work. In order to reduce the weight of the engine, electric power will be used, resulting in innovation and changes in the engine structure. As shown in Fig. 6.2.8, there is an example of the combination of no bleed air structure/engine start function.

For different flight states, a variety of devices that use similar electrical energy should be merged. For example, the B787 without bleed air can share the environmental control system ECS with the engine starter power supply PWM inverter. Since the functions and capacity of these two are basically the same, this combination of functions will have greater benefits and will allow the starter motor to no longer have an overload situation.

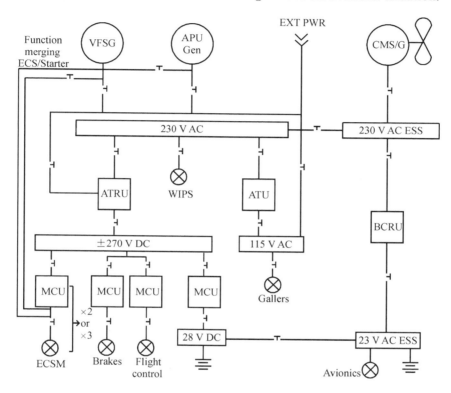

Fig. 6. 2. 8   Example of combination of bleed air structure without loop control and engine start function

According to the principle of power sharing, it is proposed to run the basic modules in parallel. For example, a three – phase PWM inverter, four inverters of 30 kV · A can use inductance so that the phases can be used alternately. Using the inductance can obtain 4 times the switching frequency, that is, the required inductance can be obtained by using 25% of the inductance of each bridge arm, and the power supply capacity of the equivalent inverter of 120 kV · A can be obtained for the compressor of the environmental control system.

The power consumption of the environmental control system is very large, most of which are powered by the variable frequency AC main generator, and ±270 V high-voltage DC power supply is used for the motors that need to be adjusted in speed. The main power supply conditions of the environmental control system are listed in Tab. 6. 2. 8.

Tab. 6. 2. 8　Main Power Supply Conditions of the environmental control system (B787)

SN	Load type	Power supply	Remark
1	Environmental control system electric fan	230 V AC	High power equipment
2	Cooling fan, cabin heating	115 V AC	More than 10 A equipment
3	Electric environmental control system compressor	±270 V DC	High power speed regulating motor

There are many motors to be used in the electric ECS, some of which can be used as asynchronous motors. Those motors are simple in structure, reliable and easy to start. Tab. 6. 2. 9 lists the AC motors for aircraft environmental control.

Tab. 6. 2. 9　AC motors for aircraft environmental control

Item	Device name	Power supply	Motor type	Quantity
1	Dryer electric fan	115 V AC,400 Hz	Induction motor	2
2	Dryer drive motor	115 V AC,400 Hz	Induction motor	2
3	Bathroom/kitchen ventilation fans	230 V　VFAC	Induction motor	—
4	Power electronics liquid cooling pumps	230 V　VFAC	Induction motor	4
5	Kitchen cooling unit fan	230 V　VFAC	Induction motor	7
6	Auxiliary cooling unit compressor	230 V　VFAC	Induction motor	4
7	Combined cooling system pump assembly	230 V　VFAC	Induction motor	2
8	Front equipment cooling supply air fan	230 V　VFAC	Induction motor	2
9	Front equipment cooling supply air fan	230 V　VFAC	Induction motor	1
10	Rear equipment cooling supply air fan	230 V　VFAC	Induction motor	2
11	Rear equipment cooling supply air fan	230 V　VFAC	Induction motor	1
12	Other equipment cooling exhaust fan	230 V　VFAC	Induction motor	1
13	Front cargo compartment heating exhaust fan	230 V　VFAC	Induction motor	1
14	Rear cargo compartment heating fan	230 V　VFAC	Induction motor	1

The air compressor motor in the electric ECS is a speed regulating motor with high speed, high power and high power density. The motor is characterized by the application of air bearings. The air bearing suspends the rotor after the motor rotates, which not only reduces the bearing loss of the rotor, but also greatly improves the bearing life and creates good operating conditions for the high-speed motor. Common Motor Starter Controller (CMSC) is a DC/AC converter/controller with a low-pass filter LC, which ensures a wide range of motor speed regulation. Similarly, the fan motor in the ram air filter also uses an air bearing, and the air bearing has become an important part of the aviation high-speed motor.

There are also many valves in the electric environmental control system, which commonly use Brushless Direct Current Motors (BLDCM) to drive. As listed in Tab. 6. 2. 10, they are the electric valves for electric environmental control.

Tab. 6. 2. 10   Electric valves for electric environmental control

SN	Name	Number of valve	Motor type	Number of motor
1	Air crew rest compartment air supply valve	2	BLDCM	2
2	Aircrew rest compartment exhaust valve	2	BLDCM	2
3	Washroom/kitchen exhaust valve	2	BLDCM	2
4	Power electronics cooling 3-way valve	2	BLDCM	2
5	Power electronics cooling supply valve	2	BLDCM	2
6	Kitchen cooling unit switching valve	7	BLDCM	7
7	Combined cooling system three-way valve	1	BLDCM	1
8	Front equipment compartment cooling override valve	1	BLDCM	2
9	Front equipment compartment cooling exhaust valve	1	BLDCM	2
10	Rear equipment compartment cooling override valve	1	BLDCM	2
11	Rear equipment compartment cooling exhaust valve	1	BLDCM	1
12	Front cargo hold heating valve	1	BLDCM	1
13	Front cargo compartment heating and exhaust valve	1	BLDCM	1
14	Rear cargo compartment heating valve	1	BLDCM	1
15	Large cargo hold heating valve	1	BLDCM	1
16	Air outflow valve	2	BLDCM	4
			DCM	2

There are also electric heaters in the electric environmental control system for cargo hold heating, rest cabin heating, etc, which are listed in Tab. 6. 2. 11.

Tab. 6. 2. 11   Electric heaters of B787 electric environmental control system

SN	Name	Number	Power supply	Electric power/(V · A)
1	Large cargo hold	1	115 V AC, 3P	1 000
2	Forward cargo hold	1	115 V AC, 3P	4 000
3	Cabin crew rest	1	115 V AC, 3P	1 100
4	Pilot cockpit	1	115 V AC, 3P	2 000
5	Dryer	2	115 V AC, 3P	—

The cabin air compressor CAC motor in the electric environmental control system is a speed regulating motor with high speed, high power and high power density. It adopts the form of rotor bearing suspension, which greatly reduces the bearing loss of the rotor and prolongs the life. The converter/controller of the motor is a DC/AC converter/controller using power electronic technology, which ensures a wide range of speed regulation of the motor.

In order to improve the reliability of the environmental control system, the motor components such as air supply fans and cooling pump components in the system adopt the redundant operating mode of backup, one of which is the main device, the other is the backup device, and the air outflow valve is controlled by 3 motors. 1 of them is manually operated, and the other 2 are backup work for each other.

There are also two sets of air conditioning pack. The redundancy of key components can significantly improve the reliability of the environmental control system, which is extremely important to improve the safety of civil aviation transportation.

Air conditioning component controller, common data system, remote data concentrate, data bus, engine indication and unit warning system, air conditioning system control panel (P5 board), dryer controller, Valve Control Unit(VCU) are the main control, protection, management and display components of the electric environmental control system. In order to ensure the operation of the environmental control system under normal and abnormal conditions, to ensure the safety and comfort of the occupants, and to ensure the work of power electronic equipment and avionics and electrical equipment, and to ensure the safety of the aircraft structure, it plays a key role. It is also the bottom line for safe transportation of civil aviation.

6.3 Environmental Requirements for Avionics and Electrical Equipment

6.3.1 Overview

With the development of technology, the efficiency of avionics and electrical equipment has improved, and the loss has been reduced, but in any case, the increase in the application of avionics and electrical equipment and the development of high-density digital electronic devices have increased the thermal load per unit volume of avionics and electrical equipment. So the result is an increase in the overall thermal load.

Aircraft is a device used all over the world, and the typical operating temperature range is − 40—90 ℃. Therefore, a typical air conditioning system designs the maximum temperature of the conditioning cabin to be 70 ℃, which is considered being the minimum requirement to meet the reliability of components. Semiconductor components can safely work above 100 ℃, but long-term operation at this temperature will affect the reliability of electronic products.

Avionics and electrical equipment are generally powered continuously from the start of power-on to the end of power-off, thus continuously dissipating heat. Equipment is usually installed in standard form in avionics and electrical equipment bays or cabinets. The cooling air enters the relevant area through the duct for cooling the equipment, and then is discharged outside the machine.

Compared with military aircraft, the thermal load of avionics and electrical equipment on civil aircraft is lower, and the application of fans to pump the ambient air of the cockpit to the inside of the equipment can meet the requirements, which will increase the temperature of the entire cockpit. However, due to the small thermal load of the avionics and electrical equipment, the environmental control system has sufficient capacity to maintain the cabin temperature at an acceptable level.

6.3.2 Electrical and Electronic Equipment Cooling

B787 uses air cooling to cool electronic and electrical equipment. Most of the electronic and electrical equipments of B787 are in the front and rear electronic and electrical equipment compartments. The display on the cockpit instrument panel is also cooled when it is operating, and some other equipment as well.

1. Equipment Cooling in the Front Electrical and Electronic Equipment Compartment

A lot of equipments of B787 need to be cooled, mainly including the power distribution board P300, P400, P500, P600 in the front electrical and electronic equipment compartment, the equipment on the equipment racks E1 and E2 and the display in the cockpit. In addition, there is the equipment on the E8 frame beside the front cargo door. The air supply fan and exhaust fan are powered by the main variable frequency AC power supply 230 V AC.

(1) Cooling Subsystem

Compartment has two parts, namely the air supply system and the exhaust air system.

1) Air Supply System

The air supply system consists of an air filter, two air supply fans, a cold air supply override valve, an air supply flow/temperature sensor, a smoke detector, and an oxygen supply pipeline.

The air-conditioning supply override valve consists of two valves in the pipeline, which are driven by two motors respectively. One of the valves is connected downstream of the air supply fan to control the air volume of the air supply, and the other valve is connected to the pipe connected to the aircraft body. The two valves act at the same time. When the air supply fan works, the air in the front cargo compartment is sucked in and filtered, and the cooling air is sent to the front electronic and electrical equipment compartment, the cockpit and the equipment on the E8 frame through the override valve.

2) Exhaust System

The electronic and electrical equipment cooling system has a set of exhaust system, which consists of exhaust fan, outside exhaust valve and pipeline. When the exhaust fan is operating, the air trapped in the upper ventilation hood of the equipment is exhausted, and is sent to the Outflow Valve (OFV) on the aircraft fuselage through the outside exhaust valve and pipeline to be discharged outside the aircraft. Another duct downstream of the exhaust fan sends air to the front cargo compartment heating line.

(2) Operating Mode of the Cooling System

There are four abnormal operating modes in the cooling system of electronic and electrical equipment, namely shutdown mode, rotor acceleration mode, override mode and override/outboard mode.

1) Shutdown Mode

The shutdown mode occurs when the air supply fans in the front and rear electrical and electronic equipment compartments lose power or the fans are faulty. The override valve and the exhaust valve should be closed, and the exhaust fan should be turned on to allow the outside air to enter the front electrical and electronic equipment compartment to cool the equipment. The air is then expelled to the front cargo compartment via an exhaust fan.

2) Rotor Acceleration Mode

The rotor acceleration mode is used when the single engine of the aircraft takes off and the CAC is not operating, or when the Air Valve System (AVS) is operating. When the air supply fan and the exhaust fan are operating, the exhaust valve opens to allow air to flow in the electrical and electronic equipment.

3) Override Mode

When the internal pressure of the aircraft is greater than the atmospheric pressure outside the aircraft, and there is smoke in the cooling air or a fire alarm in the front cargo compartment (or when the air supply fan is faulty, or the temperature of the cooling air is too high), the air supply fan and exhaust fan are used. Then the air supply fan and exhaust fan are closed, the exhaust valve is closed, and the override valve makes the air in the front electrical and electronic equipment compartment exhausted from the override valve to the outside of the machine.

4) Override/Outboard Mode

When the aircraft is on the ground or when the air pressure difference between inside and outside the cabin is small, use the override/outboard mode if there is smoke.

When operating in the override/outboard mode, the air supply fan should be turned off, the exhaust fan and exhaust valve should be opened, and the override valve should be in the override position, so that the outside air can enter the electrical and electronic equipment compartment through the override valve, and the exhaust fan shall be used after cooling the equipment.

Regardless of the normal or abnormal operating mode, it is controlled by the CCS through the signal of the sensor. Tab. 6. 3. 1 lists the abnormal operating modes of the equipment cooling subsystem in the Front Electrical and Electronic Equipment Compartment (FEEEC).

2. Equipment Cooling in the Rear Electrical and Electronic Equipment Compartment

The equipment cooling subsystem of the rear electrical and electronic equipment compartment is located in the rear cargo compartment, mainly including air supply fans, override valves, exhaust fans, exhaust valves and rear cargo compartment heating valves, etc. The cooled equipment is in the rear electrical and electronic equipment compartment. Liquid cooling pumps for power electronics that need to be cooled are located in the left and right landing gear nacelles. One end of the override valve pipeline and one end of the exhaust valve pipeline are connected to the fuselage of the aircraft, so as to communicate with the surrounding air. As listed in Tab. 6.3.2, it is the abnormal mode of the equipment cooling subsystem in the rear electrical equipment compartment, in which the outboard mode has a higher priority than the override/extravehicular mode of front electrical and electronic equipment compartment.

Tab. 6.3.1 Abnormal operation modes of the equipment cooling subsystem in FEEEC

SN	Modal name	Air supply fan	Override valve	Exhaust fan	Vent	Front cargo compartment heating air supply valve
1	Off state	Turn on	Disconnect	Turn on	Disconnect	Turn on
2	Rotor acceleration	Turn on	Disconnect	Turn on	Turn on	Disconnect
3	Override	Disconnect	Turn on	Disconnect	Disconnect	Disconnect
4	Override/Outboard	Disconnect	Turn on	Turn on	Turn on	Disconnect

Tab. 6.3.2 Abnormal modes of the equipment cooling subsystem in the rear electrical equipment compartment

SN	Modal name	Air supply fan	Override valve	Exhaust fan	Vent	Front cargo compartment heating air supply valve
1	Off state	Turn on	Disconnect	Turn on	Disconnect	Turn on
2	Rotor acceleration	Turn on	Disconnect	Turn on	Turn on	Disconnect
3	Override	Disconnect	Turn on	Disconnect	Disconnect	Disconnect
4	Override/Outboard	Disconnect	Turn on	Turn on	Turn on	Disconnect

6.4 Summary

Compared with conventional aircraft, the environmental control system in electric aircraft has undergone great changes in terms of system, structure, and control methods. The more electronic environmental control system no longer needs to bleed air from the engine, which greatly improves the performance of the engine and the whole machine.

The environmental control system of the MEA greatly increases the utilization of electrical energy, mainly the cooling of the compressor, various valves, cooling fans and electronic and electrical equipment, which puts forward new requirements for electrical technology.

6.5　Exercises

1. What are the effects of the high-altitude environment on human physiology?

2. What are the characteristics of the high-altitude working environment?

3. What are the components of the environmental control system?

4. What are the characteristics of cooling in avionics?

5. Compare the main characteristics of the environmental control system of traditional aircraft with that of MEA.

Chapter 7　Electrical Lighting Technology of MEA

Aircraft lighting mainly has two categories: internal lighting and external lighting. Internal lighting is mainly for flight crew lighting and passenger cabin service lighting; external lighting is mainly used for aircraft landing, safety, identification, taxiing, cargo handling, etc.

With the development of electrical and electronic science and technology, new requirements have been put forward for the comfort of lighting, energy-saving lighting, and high power density of the driving device. With the rapid development of Light-Emitting Diode (LED) lighting technology in recent years, it has also been commonly used in aircraft. As LED lights are constant current driven lighting devices with high luminous efficiency and good energy efficiency, they were successfully used in MEA when their drive technology was developed.

7.1　Overview

The lighting on the B787 is all based on products with LED lighting technology as the technical background. General lighting is mainly used for aircraft position marker lights and collision avoidance lights. High-Intensity Discharge (HID) lighting systems are used for landing lights, runway closure lights, wing strobe lights, airline marker lights, external cargo lights, etc.

Incandescent lighting(白炽灯) on board are used to illuminate general areas of the aircraft, such as the landing gear compartment, Nose Landing Gear(NLG) compartment and Main Landing Gear(MLG) compartment; the lighting of the Environmental Control System (ECS) compartment (Right and left); the lighting of the electrical equipment compartment at the front and rear of the fuselage and the APU and after lighting.

7.1.1　LED Lighting Technology

The B787 uses high brightness HID and LED lighting technology and is quite sophisticated, mainly using solid state technology, no moving parts and no filament fusing problems, no sudden failures.

LED lighting technology has more than 10 times the operating time of conventional lighting; no filters are needed if colour lighting is required: The colour is produced directly by the light-emitting element, so there is no loss of light. LED lighting makes it easy to achieve directional lighting over a range of viewing angles, increasing the efficiency of the lighting, unlike conventional lighting where energy is wasted. The amount of electrical energy required is therefore also significantly reduced.

7.1.2　LED Lighting Technology on the B787

The cockpit lighting for the B787 is shown in Fig. 7.1.1. The cockpit lighting provides control and indication of the work area for the crew. The cockpit lighting has a solid-state LED area light source, a brightness control panel, Common Computing Resource (CCR) software for controlling most lights and a power control unit.

Fig. 7.1.1　Cockpit lighting for the B787

The external lighting system of the B787 also uses LED technology, as shown in Fig. 7.1.2. It is a schematic diagram of the external lighting of the B787, from which it can be seen that its marker lights and anti-collision lights all use LED lighting technology.

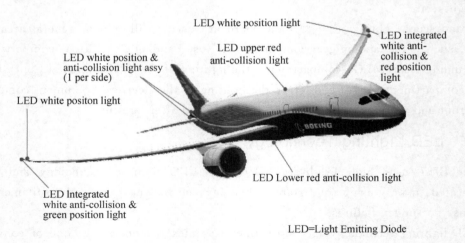

Fig. 7.1.2　External lighting layout

Fig. 7.1.3 shows high-performance LED lighting, often used for sign lights and anti-collision lights.

The MEA uses a series of high-performance LED lights made by Honeywell. These LED lights have low power consumption and higher reliability, with an estimated 20 000 hours of operating life.

Fig. 7.1.3   High performance LED lighting(Honeywell product)

The position marker signals of the B787 are shown in Fig. 7.1.4, with red lights on the left and green lights on the right, which cover an angle of 0 to 110° horizontally and −90° to +90° vertically. The lighting at the rear of the fuselage is white and covers an angle of 140° ±70° horizontally and −90°—90° vertically. When flying at night, one aircraft indicates the information about the aircraft's direction of flight to another aircraft with three colour signals: red, green and white.

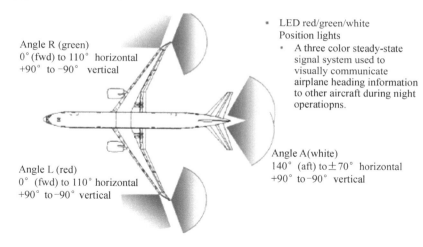

Fig. 7.1.4   B787 position sign signal light

The red anti-collision strobe light is shown in Fig. 7.1.5. The strobe light is located above and below the body. Fig. 7.1.5(a) shows a top view of the anti-collision strobe light and Fig. 7.1.5(b) shows a side view of the strobe. It has 360 ℃ overage horizontally and 0—75 ℃ overage vertically.

As shown in Fig. 7.1.6, it is a schematic diagram of the LED white strobe anti-collision lights, which are installed on the front left and right wings of the aircraft and on the tail of the aircraft. The horizontal forward coverage angle is 0 to 110°and the vertical coverage angle is −75°to 75°. There is also a white strobe anti-collision light at the rear of the aircraft with a horizontal rear coverage angle of 140°±70°and a vertical coverage angle of −75°to 75°.

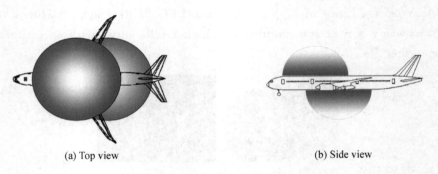

(a) Top view   (b) Side view

Fig. 7. 1. 5   Red anti-collision strobe light

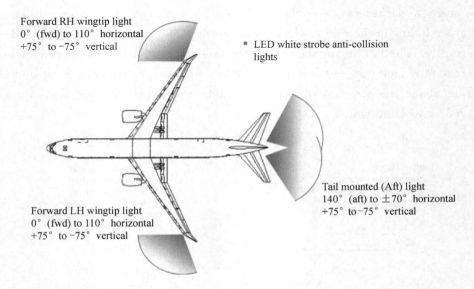

Forward RH wingtip light
0° (fwd) to 110° horizontal
+75° to -75° vertical

• LED white strobe anti-collision lights

Forward LH wingtip light
0° (fwd) to 110° horizontal
+75° to -75° vertical

Tail mounted (Aft) light
140° (aft) to ±70° horizontal
+75° to -75° vertical

Fig. 7. 1. 6   LED white strobe anti-collision light

As shown in Fig. 7. 1. 7, the LED illumination is mounted on the wing tip of the aircraft, and the upper layer of the skin is removed to reveal the LED white anti-collision light assembly and the LED red position marker light assembly.

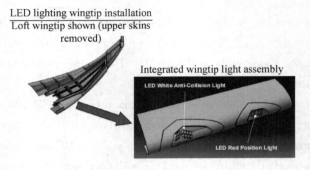

LED lighting wingtip installation
Loft wingtip shown (upper skins removed)

Integrated wingtip light assembly

LED White Anti-Collision Light

LED Red Position Light

Fig. 7. 1. 7   Wingtip mounting position for LED lighting (left wingtip, top layer of skin removed)

7.2 HID High Brightness Lighting Systems

Aircraft requires high brightness lighting mainly for landing lights, runway lights, taxiing lights, wing lights, marker lights, external cargo lights, etc.

The implementation of HID lighting technology increases reliability, mainly in the following:

① The use of xenon gas discharge in place of gradually lit incandescent filaments to produce light for HID lighting.

② The lower fuselage lighting, thanks to HID technology, has a power loss equivalent to 20% of the power of an incandescent lamp.

③ With HID lighting technology, it is at least 10 times more reliable than incandescent lighting.

HID lighting has been in production since 1998 and has been used successfully in aviation due to its energy efficiency. Fig. 7. 2. 1 shows the key components of HID high-brightness lighting, consisting mainly of the light-emitting component, the control unit and the xenon lamp.

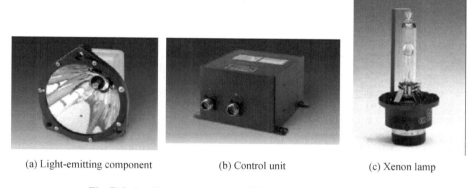

(a) Light-emitting component   (b) Control unit   (c) Xenon lamp

Fig. 7. 2. 1 Key components of HID high-brightness lighting

As shown in Fig. 7. 2. 2, it is a schematic diagram of the HID lighting technology used on the outside of the B787, mainly landing lights, runway turn signals, cargo hold lights, wing lights and marker lights.

As shown in Fig. 7. 2. 3, they are HID landing lights, 6 in total, needed to illuminate the distance to the high beam. Two 50W HID landing lights are mounted at the head of the wing and two 50W HID lights at the root of each side of the wing.

Fig. 7. 2. 4 shows the runway lighting, which requires sufficient illumination view width in addition to sufficient high beam. There is one 50 W HID light source on each side of the wing for the aircraft runway take off lights, as shown in Fig. 7. 2. 5. There are also two 50 W HID lighting sources on each side mounted in the nose for the aircraft taxiing, as shown in Fig. 7. 2. 6.

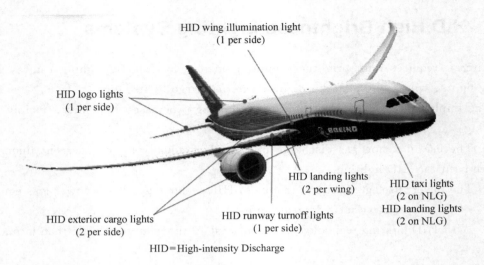

HID wing illumination light
(1 per side)

HID logo lights
(1 per side)

HID landing lights
(2 per wing)

HID taxi lights
(2 on NLG)
HID landing lights
(2 on NLG)

HID exterior cargo lights
(2 per side)

HID runway turnoff lights
(1 per side)

HID=High-intensity Discharge

Fig. 7. 2. 2　HID lighting technology for aircraft exterior lighting（B787）

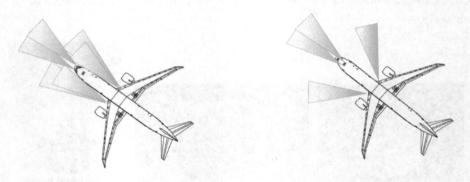

Fig. 7. 2. 3　HID landing lights (6 in total)　　**Fig. 7. 2. 4　HID runway lights (taxiing lights)**

Fig. 7. 2. 5　HID landing and taxiing lights (mounted on the wing)

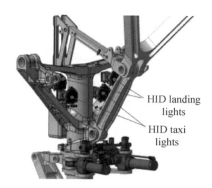

Fig. 7. 2. 6 HID landing and taxiing lights mounted on the nose landing gear

Fig. 7. 2. 7 shows a schematic of the HID wing lighting with 150 W HID light on each side. Fig. 7. 2. 8 shows HID external cargo bay lights with 50 W HID lights close to each cargo bay door, with 250 W HID lights close to the large cargo bay door and the airline logo marker lights shown with 150 W HID light on each side.

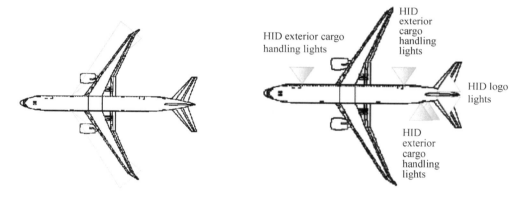

Fig. 7. 2. 7 HID wing lighting   **Fig. 7. 2. 8 HID external cargo hold lighting and marker lights**

7.3 Control of Lighting Systems

The lighting displays of the lighting system are used primarily for aircraft landing, safety, identification, taxiing and cargo handling, as well as for crew and passenger area lighting. The cockpit and cabin lighting scenes for the B787 are shown in Fig. 7. 3. 1.

7.3.1 Cockpit Lighting

A diagram of the cockpit lighting environment is shown in Fig. 7. 3. 2. The lighting system has information on the individual lighting components, their mounting locations, system interfaces, and individual operations.

Fig. 7. 3. 1　Lighting for the cockpit and cabin

Fig. 7. 3. 2　Cockpit lighting environment

The cockpit lighting equipment is shown in Fig. 7. 3. 3. The cockpit lighting fixtures require reasonable control when in use. The main controls are: the display brightness of the downward light from the cockpit top light, the system signaler, the integrated panel lighting, the main light brightness, the dome floodlight, the spotlight and the various task lighting.

The panel lighting system software in the CCR cabinet in Fig. 7. 3. 3 receives data from various control points, lighting data in various environments. The main data are monitoring control and light sensor input signals, combined input signals for the main luminance and control panels, generating and sending display and panel luminance commands, generating and sending semaphore test commands and luminance test commands.

1. Lighting Regulation

There are a number of illuminated operating buttons on the cockpit headboard P5 board, as shown in Fig. 7. 3. 4, with the following main functions.

① The controller of the main brightness control system is used to regulate the lighting of all panels and to adjust the display.

② The main brightness control switch is located on the P5 panel and is controlled by

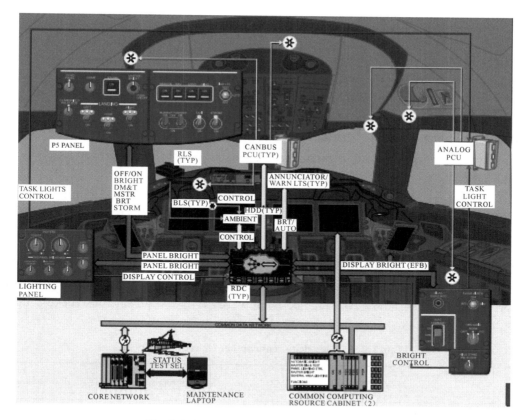

Fig. 7. 3. 3   Cockpit lighting equipment(B787)

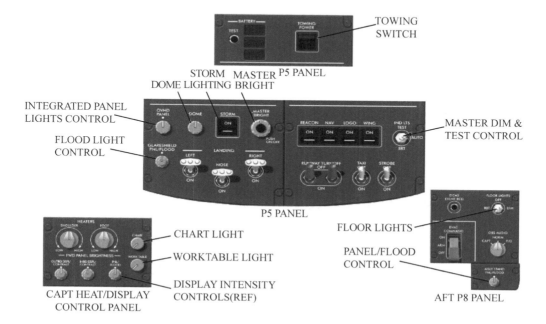

Fig. 7. 3. 4   Lighting operation buttons on the P5 board of the overhead panel (B787)

pressing the centre button, and the knob is used to adjust the brightness.

③ The digital switch data is fed into the CCR via the Common Data Network (CDN) for panel lighting control.

④ The main brightness control data is mainly available in up or down direction, panel and Electronic Flight Bag (EFB) display.

⑤ Brightness control of individual control panels, i. e. adjustment of panel indicators and displays, is possible when the main brightness control unit is in use.

⑥ The main brightness adjustment switch does not control the adjustment of the dome or flood lights.

⑦ The Master Minimum Equipment List(MMEL) allows the aircraft to send a failure message with the master brightness control as long as the switch remains off.

⑧ There are no status messages that can be displayed for failure.

2. Electricity for Lighting Equipment

As listed in Tab. 7.3.1 for an aircraft cockpit ambient lighting and instrument lighting, it is preferable to use a low voltage DC supply of 28 V to power the cockpit lighting equipment, considering that 36 V is a safe electrical voltage to use.

Tab. 7.3.1　Electricity consumption for cockpit lighting

Order number	Main lighting equipment	Lighting voltage/V	Power sinks
1	Main flight instrument panel (captain's position)	28	DC BUS 1
2	Main flight instrument panel (flight/manoeuvring position)	28	DC BUS 2
3	Power supply for ambient lighting (left and right)	28	DC ESS BUS
4	Power supply for ambient lighting (rear)	28	DC Gen. S BUS

7.3.2　Passenger Compartment

The cabin lighting for the B787 is shown in Fig. 7.3.5. The cabin conventional lighting system provides illumination for the passenger cabin seats, aisles and work areas.

Fig. 7.3.5　B787 cabin lighting

The B787 passenger cabin lights and passenger call lights and their control systems are shown in Fig. 7. 3. 6. The cabin general lighting system is illuminated by LED strips of varying lengths. The areas of the general lighting system are: passenger loading lighting, crew operating lighting and entrance lighting. The passenger cabin lighting has overhead lights and side wall coloured lights. The software in the Cabin Service System Controller (CSSC) provides lighting controls and options such as number of zones to be lit, aircraft configuration, colour or atmosphere. The general lighting system is controlled using the Cabin Crew Switch Panel(CCSP) and Cabin Crew Panel (CCP). Some components come with a remote power supply.

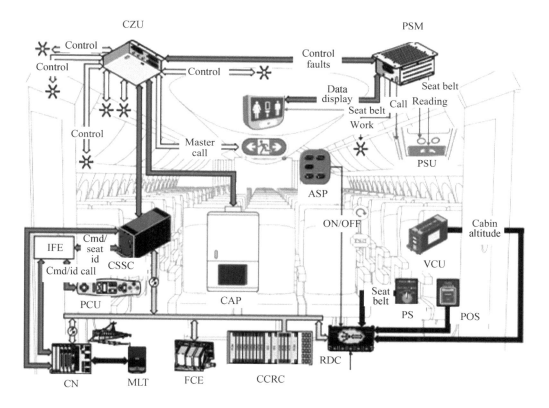

Fig. 7. 3. 6   B787 passenger cabin lights and passenger call lights and their Control Systems

Fig. 7. 3. 6 shows the Cabin Zone Unit (CZU), Passenger Service Module (PSM), Attendant Switch Panel (ASP), Passenger Service Unit (PSU), in flight entertainment, Cabin Services System Controller (CSSC), and Valve Control Unit (VCU), Passenger Control Unit (PCU), Cabin Attendant Panel (CAP), Passenger Signs (PS), Passenger Oxygen System (POS), Core Network (CN), Flight Control Electronics (FCE), Common Computing Resource Cabinet (CCRC), Remote Data Concentrate (RDC), etc. According to Fig. 7. 3. 6, the cockpit passenger lighting and call system adopts a computer-based data exchange system, and the information interactions are listed in Tab. 7. 3. 2.

Tab. 7. 3. 2 Main information interactions between passenger lighting and call systems

SN	Name	Information interaction	Object
1	Cabin Zone Unit (CZU)	Control signals	Lighting Control (LC)
		Control/Fault information	Passenger Service Module (PSM)
			Cockpit Service System Controller (CSSC)
		Master call	Cockpit Service Panel (CSP)
			Emergency Access (EA)
2	Passenger Service Module (PSM)	Control/Fault	Cabin Zone Unit (CZU)
		Seat belt, reading	Passenger Service Unit (PSU)
		Data display	Passenger Call (PC)
3	Cockpit Service System Controller(CSSC)	Control/Fault information	Cabin Zone Unit (CZU)
		Common seat ID	In-Flight Entertainment System (IFES)
		Two-way data interaction	Common Data Network (CDN)
		Two-way data interaction	Core Network (CN)
4	Remote Data Concentrate (RDC)	Two-way data interaction	Common Data Network (CDN)
		Seat Belts	Passenger Sign (PS)
		DEPLOY	Passenger Oxygen System (POS)
		ON/OFF	Cabin Crew Switch Panel (ASP)
		Cabin Altitude	Valve Control Unit (VCU)
		Flight Mode	Flight Deck Door/Engine Running (FDD/ER)
		Airplane State	Landing Gear Lever/Flap Lever (LGL/FL)
		Over Temp	Environmental Control System (ECS)

7.3.3 Lighting Control Solutions

Lighting control schemes for lighting systems are usually of two types, namely single-channel control switches and internal clock switches with software. A single-channel control switch is shown in Fig. 7. 3. 7.

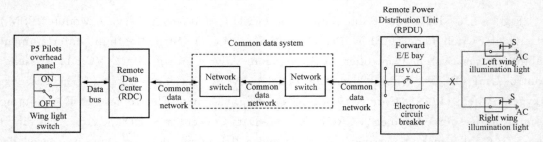

Fig. 7. 3. 7 Single-channel control switch

The main switch for the wing lights is located in the cockpit on the overhead panel P5, which communicates with the RDC via the data bus, and with the common data centre system via the CDN bus, and then via the CDN with the electronic circuit in the front electronics compartment. The circuit breaker is connected to the electronic circuit in the front electronics compartment, which is controlled by the RPDU to supply power to the left and right wing lights, which are powered by 115 V AC in the diagram.

The software switch control system with an internal clock, as shown in Fig. 7.3.8, is similar to the single-channel switch control except that a common data system is used for data distribution and control, which will not be repeated.

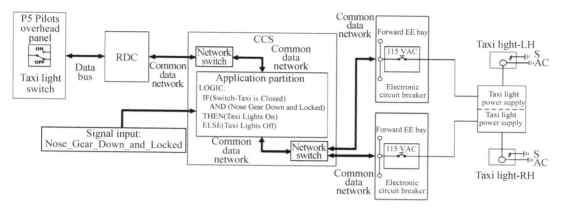

Fig. 7.3.8 Software switch control mode with an internal clock

7.4 Sequence for the Normal Use of Aircraft Exterior Lights

① After the aircraft has been switched on to the main power switch, the pre-flight mechanic turns on-navigation lights, wing lights and LOGO lights as required.

② Turning on the red anti-collision light when the aircraft is launched (also for on ground trial run).

③ The aircraft starts its engines and then turns on its turn signal in preparation for taxiing out.

④ Once permission to slide out has been obtained, turn on the slide light and start sliding.

⑤ Turning on the white anti-collision lights upon entering the runway.

⑥ Turning on the landing lights for takeoff once clearance has been obtained.

⑦ After leaving the ground, turn off the glide light and turn light (can be turned off automatically after retracting the landing gear).

⑧ Turning off the white flash when the height rises above 10 000 feet.

⑨ At a minimum, the red flashing lights, navigation lights should be kept on during cruise, with the LOGO and wing lights on as required.

⑩ Aircraft descending to 10000 below feet with white flashing lights on.

⑪ Turning on the taxiing lights after the aircraft has been lowered on the landing gear.

⑫ Switching on the landing lights during the final approach phase.

⑬ After grounding, turn on the turning lights, turn off the landing lights and turn off the white flashing lights.

⑭ Turning off the glide lights, red flashing lights after gliding into position.

⑮ If the aircraft is no longer scheduled for flights and stay overnight, the aircraft will be turned off by the post flight crew after the final switch off of the navigation lights.

7.5　Summary

This chapter describes the electrical lighting technology on board MEA, mainly using energy efficient devices based on LED, the different power supply technology for LED lights compared to incandescent lights and the requirements for energy saving technology, miniaturization and reliability.

In addition to illumination, lighting also serves the purpose of information transmission and display, and requires sufficient electrical power supply, so energy-efficient aviation also requires its highly efficient power supply and efficient luminescence.

7.6　Exercises

1. What are the basic requirements for the aircraft lighting system?
2. What are the technical characteristics of high-brightness lighting technology?
3. Why does lighting usually use low voltage DC power?
4. What are the two usual lighting control schemes for the lighting system?
5. What should be paid attention to during the use of aircraft external lighting?

Chapter 8　Brake System of MEA

8.1　Introduction

The landing gear is used to support the weight of the aircraft and absorb the impact energy when the aircraft is parked on the ground, taxiing and taking off and landing. Modern aircraft, with the exception of a few small ones, have their landing gear built into the fuselage or wings after takeoff. Landing gear accounts for about 2.5%—4% of the weight of the aircraft.

Landing gear, according to the position and layout of the aircraft from the center of gravity, is divided into the front three-point landing gear and the rear three-point landing gear, the rear three-point landing gear is lighter than the front three-point landing gear, the brake is too fierce when there is a danger of "taking the top", poor stability when sliding, has been eliminated. The most commonly used landing gear on aircraft is the front three-point landing gear, with the nose wheels under the nose, away from the center of gravity, and the two main wheels positioned symmetrically left and right behind the center of gravity. This configuration, when landing at higher speeds, uses hard brakes and does not stand upside down. During landing and taxiing, the pilot has a wide field of vision, which can prevent the plane from turning in taxiing.

As shown in Fig. 8.1.1, it is the brake shape diagram of the landing gear of the B787 aircraft, using the front three-point landing gear, with the front landing gear and the left and right rear main landing gear. Weight bearing support and glide phase plane direction change of the role of the distribution in left and right sides of the aircraft landing gear after, in the later of centre of gravity of the aircraft fuselage, bearing the support of the plane, is the weight of the role, the left and right sides of the landing gear respectively have 4 tires, so the whole plane has a total of 10 tires, the work of each tire must be controlled and coordinated.

During the landing process of the plane, when the wheel is ground instantly or runs on an uneven runway, the plane will hit the ground violently, only inflatable tires can play a few buffer, the main impact energy relies on shock absorbed to absorb. Shock columns are self-enclosed hydraulic devices that support the aircraft, absorbing and mitigating the huge impact loads which generated during landing to protect the aircraft structure.

When the shock absorbed is compressed by impact, the air acts like a spring, storing energy. In addition, the oil travels through the hole at great speed, absorbing much of the impact energy and converting it into heat, keep the plane soon stabilized after landing. The retractable power source of the landing gear is hydraulic or pneumatic, and its operation is

Fig. 8. 1. 1　Outline of B787 landing gear brake

realized by an electrical control device.

With the development of aviation technology and electrical technology, the landing gear and its braking system have been developed, especially the development of high-power technology and electronic technology, which promotes the application of electric brake actuation technology, sensor technology and information processing technology in the braking system. The following takes the B787 as an example to introduce.

8.2　Landing Gear Braking System

Aircraft landing gear braking system is very complex, the main components are Main Landing Gear(MLG), Nose Landing Gear(NLG), brake controller, brake actuator, wheel speed measurement sensor, tire pressure measurement sensor, brake temperature sensor, brake pedal sensor, handbrake, remote data communication network, electric brake power supply, brake indicator light, ground service, etc.

8.2.1　Basic Composition of Landing Gear Braking System

1. Basic Composition

B787 landing gear brake system is mainly composed of brake actuation system, brake system control unit, Electric Brake Actuator Controller (EBAC) and sensor, etc. As shown in Fig. 8. 2. 1, it is the composition principle of MEA B787 brake actuation system.

The key components in Fig. 8. 2. 1 are pedals, nose wheel idle speed (automatic brake), nose wheel brake light, automatic brake selector switch, left brake system control unit, left MLG pedal brake control, outer channel automatic brake control, parking brake control, landing gear retraction brake control, left wheel anti-skid hydraulic control, right brake system control unit, right brake system control unit, right main landing gear pedal brake control, Electric-Brake Power Supply Unit(E-BPSU), MLG axle RDC, anti-skid function, brake actuation/position/load, RDC, left landing gear lever up, left and right tires in

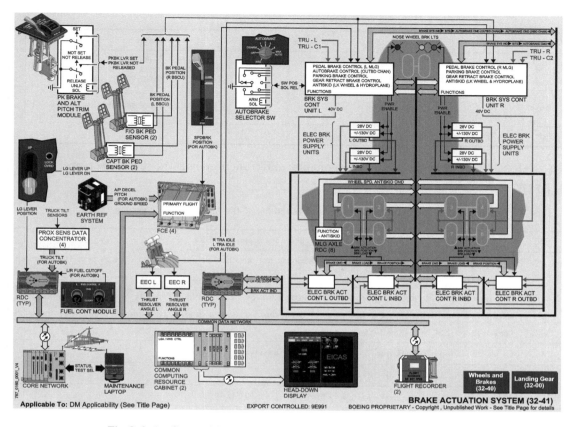

Fig. 8. 2. 1 Composition principle of MEA B787 brake actuation system

neutral, left and right fuel cut off, EBAC on outer left, electric brake on inner right Brake actuation controller, EBAC on the right outer side, etc. Taking the landing gear system of B787 as an example, there are 17 components or assembles. Tab. 8. 2. 1 lists the main components and parts of the landing gear system of B787.

Tab. 8. 2. 1 The main components and parts of the landing gear system(B787)

SN	Name	Number	SN	Name	Number
1	Electric Brake Actuator Controller(EBAC)/n	4	7	Parking brake and spare pitch adjustment module/set	1
2	Brake pedal sensor (left, right)/n	4	8	Main landing gear wheels and tires/sets	8
3	Brake System Control Unit(BSCU)/n	2	9	Nose landing gear wheels and tires/sets	2
4	Main Landing Gear (MLG) Remote Data Controller (RDC)/n	8	10	Nose wheel rotary brake/n	2
5	Automatic brake selector switch ASS/n	1	11	Tire expansion valve/n	10
6	Magnetic ring/n	8	12	Main landing gear brake/n	8

(Continued)

SN	Name	Number	SN	Name	Number
13	Electric Brake Actuator (EBA) (on respective main gear brakes)/each	4	16	Remote data concentrate/on nose gear shaft	2
14	EBA transmission (on respective main gear brakes)/n	4	17	Tire pressure sensor/n	10
15	Brake temperature sensor/n	8	—	—	—

Tab. 8. 2. 2 lists the main components and functions of the brake actuator system of B787.

Tab. 8. 2. 2   The main components and functions of the brake actuator system (B787)

SN	Main components and functions
1	Left and right BSCU, contains all control functions, but does not include Common Core System (CCS) about brake control functions
2	The input signal of the aircraft cockpit is directly connected to the BSCU, mainly including the brake pedal sensor, parking brake and spare spacing module parking brake joystick switch, automatic brake selector switch (only the left BSCU has) and landing gear handle
3	Other BSCU input signals from CCS's Common Data Network (CDN)
4	Left and right BSCU provide data for 4 EBACs
5	The power supply of the brake system is provided by 4 EBPSUs
6	Power supply of BSCU and EBPSU: provided by Transformer Rectifier Unit (TRU) during normal operation, main battery backup in emergency, or tow power supply
7	Each EBAC receives a power supply from one of the 4 electric brake power supplies EBPCU
8	Each Electric Brake Actuator Controller (EBAC) can power 8 EBACs per pair
9	Electric Brake Actuator (EBA) provides the necessary force to the brake disc
10	The main landing gear hub has the main landing gear axle Remote Data Concentrate (RDC), the main functions are: monitoring wheel speed, providing main anti-skid commands to the relevant BSCU, monitoring brake temperature and monitoring tire pressure
11	The remote data controller RDC on each Nose Landing Gear (NLG) hub only monitors the pressure of each tire

As listed in Tab. 8. 2. 3, it is a list of wheel and brake system assembly interfaces of B787.

Tab. 8. 2. 3   Wheel and brake system assembly interfaces (B787)

SN	Name	SN	Name
1	Electric brake power supply unit	4	Flight control electronics
2	Nose landing status light	5	Electronic and Electrical Controller (EEC)
3	Landing gear control system	6	CCS from RDC

The wheel and brake systems have functional interfaces to the CCS, namely Landing Gear Actuation(LGA) and Nose Wheel Steering (NWS) control functions as well as display and crew warning functions.

2. Brake System Control Unit

(1) Function

According to Fig. 8. 2. 1, each Brake System Control Unit (BSCU) has two channels, namely the main landing inner tire and the outer tire control channel. Each channel provides data to other BSCU channels through CDN. Data from other systems is sent to the BSCU on the CDN. The BSCU provides the EBAC drive commands on the controller (CAN) bus. The BSCU controls the operation of all brake systems, of which the left BSCU controls the 4 brakes on the left MLG, the right BSCU controls the 4 brakes on the right main landing gear MLG, each BSCU channel has the same loadable software, and each BSCU channel has 4 functions, as listed in Tab. 8. 2. 4.

Tab. 8. 2. 4   Functions of BSCU

SN	Function
1	Receives all input signals for brake operation (Auto brake input only goes to the left outboard channel)
2	Enables the EBPCU to power the corresponding EBAC
3	Supply power to the Remote Data Concentrate (RDC) of the corresponding brake axis
4	Send the brake command data to the corresponding EBAC to adjust the Electric Brake Actuator (EBA) to suit the necessary braking power

The BSCU on the left and right uses the manual brake with the input signal from the brake pedal sensor, where the stop brake lever operates from the output angle of the left and right thrust analyzer of the Electronic and Electrical Controller (EEC) on the CDN.

(2) Power Supply

The brake power supply is powered by the enabling trigger signal issued by the BSCU. The BSCU on the left and right side provide the power supply trigger signal to the respective internal and outer power supplies respectively, and the four electric brake power supply units provide ± 130 V DC power supply to the respective electric brake actuator controllers for brake operation.

The transformer rectifier unit supplies power to each BSCU channel, the left internal and outer electric brake power supply unit provides 28 V DC power supply to the left BSCU, the right brake system control unit BSCU provides 40 V DC current to the remote data concentrate at the main landing gear shaft, and the right brake system control unit also provides 28 V DC to the remote data concentrate on the forward landing gear shaft.

(3) Control Information

For all braking modes, the brake modules on the left and right sides use instructions from the CAN bus, which are the wheel speed and anti-skid instructions input from the remote data concentrate output on the eight main landing gear shafts.

As shown in Fig. 8. 2. 2 (a), it is the electric brake system control unit, located in the front electronic and electrical equipment compartment on both sides of the front wheel compartment. The BSCU on the left and right sides are respectively installed in the left and right electronic equipment module, as shown in Fig. 8. 2. 2 (b). There are access ports on both sides of the electronic and electrical equipment module.

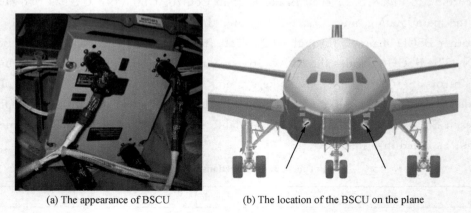

(a) The appearance of BSCU   (b) The location of the BSCU on the plane

Fig. 8. 2. 2　Electric brake system control unit

Since there is no direct data exchange between the BSCU channels, each brake system control unit sends the data to the other BSCU channels through the common core system. The BSCU also has an auxiliary anti-skid function, which can emit the braking force on the wheel, realizing the aircraft hydraulic system and its ground touch protection function (rear wheel only). When there is no input for the brake operation, the BSCU causes the electric brake actuator controller to put the electric brake actuator in an inactive state.

Tab. 8. 2. 5 lists the control mode of BSCU, each channel sends the control data to the relevant electric brake driving controller to operate the front and rear wheel brakes, and the brake system control unit has internal software that can be loaded.

Tab. 8. 2. 5　Control mode of BSCU

SN	Name	SN	Name	SN	Name
1	Alternate mode	4	Auxiliary anti-skid mode	7	Landing gear retraction mode
2	Normal braking mode	5	Park and adjust mode	8	Coasting brake mode
3	Automatic braking mode	6	Engine run stop mode	9	Test mode

BSCU has the function of data exchange between RDC and CDN, and has BSCU analog and digital signal interfaces.

3. Electric Brake Actuator Controller

(1) Composition

The electric brake actuator controller each controls the brake actuator on the main landing gear brake, and the EBA can push the brake disc to brake. According to Fig. 8. 2. 1, there are 4 electric brake driving controllers, namely Electric Brake Actuator Controller (EBAC) inside and outside the left board and EBAC inside and outside the right board. Each

EBAC has two channels, one channel controls the electric brake actuator EBA on the front wheels (1,2,3 and 4 in Fig. 8. 2. 1) and the other channel controls the EBA on the rear wheels (5,6,7 and 8 in Fig. 8. 2. 1).

Fig. 8. 2. 3 shows the B787 EBAC assembly, where Fig. 8. 2. 3 (a) shows the shape diagram of the EBAC assembly, and Fig. 8. 2. 3(b) shows its position on the aircraft. The B787 aircraft EBAC is installed in the rear electronic equipment compartment of the aircraft. There are four such controllers.

 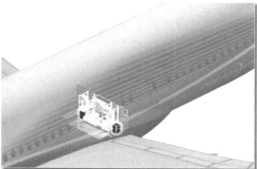

(a) Appearance of the EBAC assembly

(b) Installed in the rear electronic equipment compartment of the aircraft

Fig. 8. 2. 3　EBAC(4 in total)

The EBACs on the left and right sides respectively operate the electric brake actuator on the corresponding outer wheel, and the left and right inner EBACs respectively operate the electric brake actuator on the corresponding inner wheel. Four electric brake actuators operate together to provide the necessary braking force.

The motor is used to drive the ball screw to set the necessary braking force on the brake disc. Each force transformation generates an electrical signal. When there is no input signal, the internal friction brake stabilizes the motor shaft. There are also brakes on the electric brake assembly. Temperature sensor is used to monitor the temperature of the brake components.

When the EBAC does not input a signal to the EBA motor, the motor will decelerate and immediately turn to generator operation. Since the motor changes from the motor state to the generator state, the electrical energy is fed back to the internal capacitor in the EBAC for energy storage, which increases the response speed, and reduces the peak transient current from the E-BPSU.

As shown in Fig. 8. 2. 4, it is the outline drawing of the electric brake assembly, which is a product of GoodRich Company. The electric brake assembly consists of a carbon brake disc, an electric brake actuator and a brake cable.

The brake actuators all have electric motors, friction brake discs, solvers, ball wire bars, and load monitor components. The cable connects the actuator on each brake to the junction box, and the internal load unit measures the force of the actuator on the brake and sends the data to the corresponding EBAC. There is a Wear Pin (WP) on each side, using the special function of central maintenance calculation. Each main landing gear MLG brake weighs 246 lb (note: 1 lb＝ 0. 454 kg).

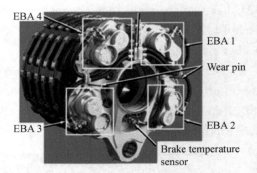

Fig. 8. 2. 4　Outline drawing of the electric brake assembly (GoodRich)

(2) Power Supply

The brake power supply is powered by the enabling trigger signal issued by the BSCU, and the BSCUs on the left and right side provide the power trigger signal to the respective internal and outer power supplies. The four electric brake power supply units provide the ±130 V DC power supply to their respective electric brake actuator controllers for brake operation.

(3) Control Information

Connect the power supply to the special electric brake actuator through a special cable, and transmit the EBA drive data to the common data center, the EBAC receives Brake Position Commands (BPC) from the corresponding inboard or outboard channel of the BSCU, each EBAC channel sends EBA force and position data back to the BSCU, and the EBAC transmits brake light and fault data directly to the CDN.

Each BSCU channel sends brake commands to the associated EBAC on a different CAN bus. EBAC only sends relevant data on the CAN bus in response to the command data from the BSCU bus. If the EBAC cannot receive the ±130 V voltage and brake force commands, the EBAC will recognize the fault. Each EBAC controls the brakes on the front/rear wheel pair and provides an analog output signal for each channel to control the operation of the 4 electric brake actuators.

Each EBAC sends brake actuator data to the CDN through the RDC, as listed in Tab. 8. 2. 6.

Tab. 8. 2. 6　Connections between RDC and CDN

SN	Name	SN	Name
1	Remote data concentrate RDC7 for inner side of left EBAC	3	Remote data concentrate RDC8 for inner side of right EBAC
2	Remote data concentrate RDC9 for left EBAC outside	4	Remote data concentrate RDC10 for right EBAC outside

The EBAC provides the data to the RDC for the cockpit display and for the onboard maintenance system. The necessary data provided to the EBAC channel via the CDN and RDC are listed in Tab. 8. 2. 7.

Tab. 8. 2. 7 The necessary data provided to the EBAC channel via the CDN and RDC

SN	Name	Function
1	ground test switch	Data loading enabled and normal usage enabled
2	Radio altimeter in Flight Control Management Cabinet(FCMC) via FCE	Test aircraft altitude
3	Proximity sensing system (air/ground sensor)	Main Landing Gear(MLG) car inclination (on the ground)
4	Brake system control unit	Normal power supply, battery emergency power supply
5	Onboard Maintenance System (OMS)	Controls the landing gear lever position

The EBA only uses the power supply when the brake is moved, and the friction brake locks the EBA when not moved. The load monitor monitors the force of EBA.

The EBA provided mechanical force to the brake discs and four EBA were assembled per brake. Each EBA has an electric motor and a gear transmission system. The EBAC sets the braking force through the transmission train to operate the motor, and the internal friction actuator will lock the motor position without moving the wheel. For energy efficiency, the EBAC provides control power to the EBA only when changing the braking force is required.

There are two brake Wear Pointers(WP) on each brake, the height of the wear pointer above the guide decreases as the brake disc wears. If the needle is at or below the specified position, the brake disc will be completely worn and must be stopped to check the height of the worn needle.

4. Sensors

The brake system mainly has four kinds of sensors, namely, wheel speed measurement magnetic ring, tire pressure sensor, point brake temperature sensor and brake pedal sensor.

(1) Magnetic Ring for Wheel Speed Measurement

As shown in Fig. 8. 2. 5, the wheel speed measurement magnetic ring sensor in each main landing gear MLG wheel and tire provides the RDC magnetic pulse on the main landing gear MLG shaft associated with the wheel speed. The RDC uses wheel speed data for anti-skid control, and the temperature sensor in each MLG brake is also connected to the RDC on the associated main landing gear MLG shaft.

As shown in Fig. 8. 2. 5, the wheel speed measurement magnetic ring has a magnetic ring sensor in each MLG wheel and tire, which can give the RDC magnetic pulse on the MLG shaft associated with the speed of the wheel. The RDC uses wheel speed data for anti-skid control, and the temperature sensor in each main landing gear brake is also connected to the RDC on the associated main landing gear shaft.

(2) Tire Pressure Sensor

As shown in Fig. 8. 2. 6, the tire pressure sensor detects the pressure in the main and the nose landing gear tire. Both have the same wireless tire pressure sensor on the wheel and can measure the pressure from 0 to 350 psi tire.

Fig. 8. 2. 5   Magnetic ring for wheel speed measurement

Usually, each tire pressure sensor has 3 channels. The RF pulse signal of RDC on the MLG or NLG axis charges the internal capacitance of the sensor, and monitors the change of the capacitance in the sensor with the tire pressure. The RDC on the axle provides brake temperature and tire pressure data to the BSCU on the CAN bus. The BSCU sends the relevant(有关的) data to the CDN for display in the cockpit.

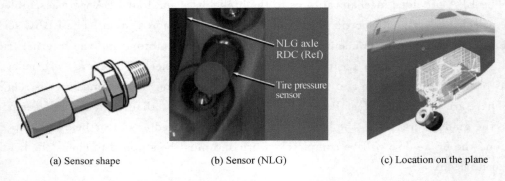

(a) Sensor shape   (b) Sensor (NLG)   (c) Location on the plane

Fig. 8. 2. 6   Tire pressure sensor

(3) Point Brake Temperature Sensor

As shown in Fig. 8. 2. 7, it is the electric brake temperature sensor. The brake temperature sensor monitors the temperature of the actuator in each MLG. Usually, the brake temperature sensor is a thermo-couple type sensor. The output signal of the temperature sensor is connected to the RDC of the main landing gear MLG, and sends the brake temperature data to the BSCU, which transmits the brake temperature data to the CDN for display in the cockpit.

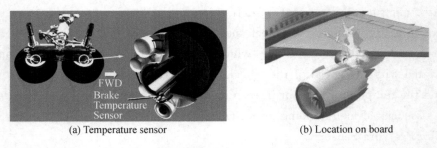

(a) Temperature sensor   (b) Location on board

Fig. 8. 2. 7   Electric brake temperature sensor (B787)

(4) Brake Pedal Sensor

Fig. 8. 2. 8 shows the actual picture of the brake pedal sensor and its position on the airplane. The brake pedal sensor outputs an analog signal proportional to the pedal force, a spring is installed on the pedal, and the brake pedal sensor moves with the pedal against the spring force. The brake pedal sensors on the left and right provide input signals to the BSCU, allowing the pedal-related differential brake action. Each brake pedal sensor has 2 Linear Variable Differential Transformers (LVDTs); the output is connected to the respective BSCU.

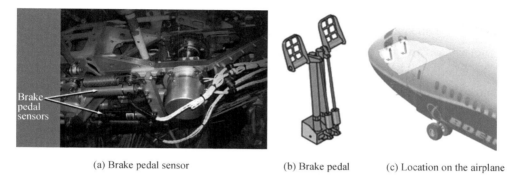

(a) Brake pedal sensor   (b) Brake pedal   (c) Location on the airplane

Fig. 8. 2. 8 Brake pedal sensor and its position on the airplane

There are two brake pedal sensors under the captain and pilot. The brake pedal sensors are located in the front compartment below the cockpit floor. These sensors are assembled with the brake pedal and work listed in Tab. 8. 2. 8.

Tab. 8. 2. 8 Operation of the brake pedal

SN	Operation
1	The signal from the brake pedal sensor is delivered to the Brake System Control Unit(BSCU)
2	The electric brake start signal of the pedal triggers the BSCU, enabling the BSCU to supply power to the electric brake actuator controller
3	BSCU sends brake command data to each EBAC
4	EBAC can adjust the proper position of electric brake
5	During coasting: (1) The front and rear wheel pairs usually have only one electric brake actuation. (2) Each brake pedal release and subsequent operation results in another brake operation on the front and rear wheel pairs. (3) There must be enough brake pedal force to make the front and rear brakes work
6	If one brake fails, the active brakes in the front and rear wheel sets will always work during coasting
7	During coasting to stop, the BSCU is in the hold mode to prevent the ever-changing signal input to the electric brake, resulting in frequent use of the pedals, and the EBA works continuously
8	Low speed, dragging and one brake pedal action at a time, other brakes operate on each front/rear wheel pair

The power busbar of the brake pedal connects the brake pedals of the captain and co-pilot together. The left and right side brake pedals connect the captain and co-pilot's brake pedals to supply power respectively through the left and right busbars, and the movement of the captain's or co-pilot's pedals will enable the BSCU of the E-BPCU to supply power to the EBAC. According to the size of the spring force, the pedal changes the position of the sensor. The BSCU uses the position signal from the sensor, and sends a braking force command to the EBAC with a signal proportional to the position signal to make the braking action.

8.2.2　Remote Data Communication Network

On the one hand, the operation of the whole aircraft wheel brake system requires the data of the wheel, and on the other hand, the operation data from the captain and the co-pilot of the cockpit, with a lot of operation modes. Therefore, it is necessary to understand the network structure of its data, as shown in Fig. 8. 2. 9, which is the remote data network structure diagram of B787.

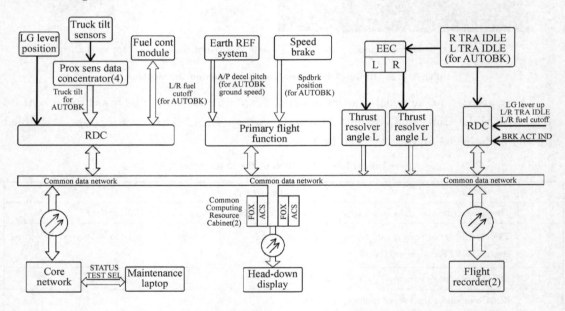

Fig. 8. 2. 9　Remote data network structure diagram of B787

As can be seen from Fig. 8. 2. 9, the data of the remote data network comes from the RDC, the flight control device, the Electronic and Electrical Controller(EEC) that gives the left and right thrust, the flight recorder, the common computing resource cabin, and other data acquisition devices.

1.　Remote Data Concentrator

The Remote Data Concentrator(RDC) is used to collect the data on the main landing gear and the nose landing gear, and transmit it to the CDN with a wireless network. There is RDC on each wheel shaft of the landing gear. Therefore, there are 10 RDCs in B787, including 2 nose landing gear and 4 on the left and right sides.

(1) The RDC on Main Landing Gear MLG Axis

Each main landing gear wheel and tire assembly has a magnetic ring and connector for measuring the aircraft wheel speed, tire pressure sensor and tire expansion valve. There is an MLG shaft RDC on each shaft hub. The main components on the MLG are listed in Tab. 8. 2. 9.

Tab. 8. 2. 9　Main components on the MLG

SN	Name	Number
1	MLG axis RDC/n	1
2	Magnetic ring and connector/set	1
3	Tire expansion valve/n	1
4	Tire pressure sensor/n	1
5	Nose landing gear brake/n	4

Fig. 8. 2. 10 shows the MLG axis RDC and its position on the aircraft.

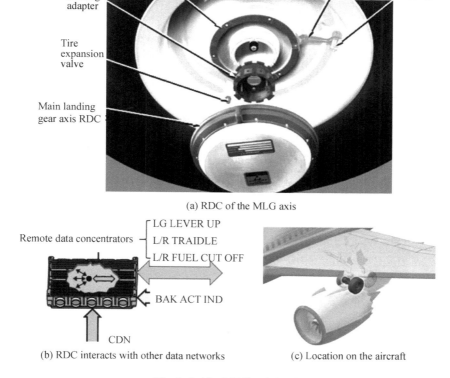

(a) RDC of the MLG axis

(b) RDC interacts with other data networks　　　(c) Location on the aircraft

Fig. 8. 2. 10　MLG axis RDC

A RDC on the main landing gear axle is used to provide raw anti-skid data, monitor tire pressure and brake temperature. Count the magnetic pulses generated by the electromagnetic coil rotating with the wheel, calculate the speed of the wheel, and calculate the magnitude of the force required for anti-skid, At the same time, the calculation data is sent to the BSCU, so that the braking command is sent to EBAC. The anti-skid function does not have to receive input commands from the cockpit, as long as it is in braking mode.

The RDC on the MLG axle transmits the transient Radio Frequency (RF) signal of the tire pressure sensor and the signal of the brake temperature sensor, so that the system receives the tire pressure data and tire temperature data. The main landing gear axis RDC performs the following operations, as listed in Tab. 8. 2. 10.

Tab. 8. 2. 10   Operations performed by the main landing gear axis RDC

SN	Name
1	Monitoring ferromagnetic coil pulses to computer wheel speed
2	Calculate the skid instruction
3	Monitor brake temperature based on brake temperature sensor data
4	Monitor tire pressure with wireless tire pressure sensor

The RDC on the MLG axle has a data connection with the BSCU to calculate the force required to prevent the wheel from slipping. The BSCU adjusts the braking force of the EBAC according to the data. The RDC on the MLG axle can load internal software.

In addition to 1 RDC, there are 16 ferromagnetic on the MLG wheel shaft. The Hall effect sensor in the RDC monitors the magnetic pulse when the wheel rotates, and the wheel speed is proportional to the magnetic pulse rate. If the wheel speed is too slow, send the anti-skid command to the corresponding BSCU on the CAN bus. If the BSCU cancels the anti-skid command from the brake drive input, the braking force will decrease to prevent the wheel from slipping. The anti-skid function can work in the braking mode.

The RDC on the MLG or NLG axle outputs radio frequency pulses to instantaneously charge the capacitor in the wireless tire pressure sensor on the wheel and output an RF signal to give the axle RDC tire pressure data.

There is also a temperature sensor on the brake connected to the RDC on the MLG axle. Both the RDC on the MLG and NLG axles can send tire pressure and brake temperature data to the BSCU on the CAN bus. Nose landing gear NLG axle RDC sends tire pressure data to the right BSCU channel on the CAN bus.

The power supply of the RDC on the left and right main landing gear shafts is 40 V DC power supply, which is provided by the respective BSCU. The operating power of the "red, yellow, blue" status lights of the NLG is 28 V DC power supply, provided by the BSCU.

When the landing gear lever is moved to the DOWN position, the BSCU will immediately disconnect the power supply of the RDC on the MLG shaft, reset the internal circuit of the RDC, and prepare for the subsequent braking operation.

(2) RDC on the NLG Axis of the Nose Landing Gear

As shown in Fig. 8. 2. 11, it is the RDC on the NLG axle of the NLG, in which Fig. 8. 2. 11(a) is an exploded view of the RDC on the axle, and Fig. 8. 2. 11(b) is the picture of the wheel hub. The RDC on the NLG axle is used to monitor the tire pressure and send a radio frequency trigger signal to the relevant tire pressure sensor to receive the tire pressure data, and the nose landing gear RDC receives the right BSCU and sends the data. NLG axle RDC monitors tire pressure via wireless tire pressure sensors. Each NLG axle RDC sends tire pressure data to the right outboard BSCU channel on the CAN bus.

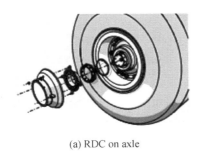

(a) RDC on axle   (b) Front hub diagram

Fig. 8. 2. 11 RDC on nose gear axle

2. Brake System Control Unit Analog and Digital Signal Interface

Each BSCU channel receives data from the RDC on the front main landing gear MLG axle and the RDC on the aft main landing gear MLG axle. The RDC on each MLG axle sends data on a different CAN bus. These data are mainly wheel speed, skid command, tire pressure, and brake temperature.

The BSCU controls the auxiliary anti-skid operation to lock the wheels for protection and prevent horizontal dragging for landing protection. Lock the wheel protection by comparing the wheel speeds of the front and rear wheels.

The brake controller has 6 kinds of analog and digital signal interfaces. As listed in Tab. 8. 2. 11, they are the analog and digital signal interfaces of the BSCU.

Tab. 8. 2. 11 Analog and digital signal interfaces of the BSCU

SN	Name	Location
1	Captain brake pedal position sensor, Linear Variable Differential Transformer (LVDT)	Left sensor in left channel, right sensor in right inner channel
2	Flight/Handling brake pedal position sensor, linear variable differential transformer	Left sensor in left outer channel, right sensor in right outer channel
3	Up and down switches in the landing gear lever module	All channels
4	Parking brake lever switch contacts and unlock solenoid	Left and right outer channels
5	Automatic brake selector switch contactor and Solenoid Valve(SV)	Left lateral channel
6	Thrust Resolver Angle(TRA) idle thrust control module switch	Left medial and lateral channels
7	Nose gear status indicator	Right medial channel

Locked wheel protection is performed by comparing the wheel speeds of the front and rear wheels on the wheel set. If the speed of the front and rear of the wheel is 30% slower than that of the other wheel, the corresponding BSCU will give the brake signal on the slower wheel to protect the wheel. Wheel speed with a relative speed to the ground less than 25 knots is prohibited (note: 1 knot=1.852 km/h).

3. Data Interaction Between RDC and CDN

Each BSCU has 2 brake power supply control units BPCU, each BPCU contains hardware and software interfaces for brake control functions. The 2 channels of the BSCU output the inner and outer brake control data respectively. The output data of the BSCU is connected to the CDN through the RDC, Tab. 8.2.12 lists the data interaction between RDC and CDN.

Tab. 8.2.12　Data interaction between RDC and CDN

SN	Name	SN	Name
1	RDC 1 for inner side of left BSCU	3	RDC 2 for inboard channel of right BSCU
2	RDC 3 for outboard channel of left BSCU	4	RDC 6 for outboard of right BSCU

There is no direct data connection between the BSCUs. The data provided by the BSCU to the RDC is obtained from other BSCU data available through the CDN, which can be used for cabin instructions and onboard equipment maintenance.

The CDN can also provide data to RDC, and further provide data to BSCU to realize some necessary functions or operations. The data provided by CDN to RDC is listed in Tab. 8.2.13.

Tab. 8.2.13　Data provided by CDN to RDC

SN	Name	Function
1	Ground test switch	Data loading enabled and normal usage enabled
2	Earth Reference System (ERS)	The Flight Control Module (FCM) in the Flight Control Electronics (FCE) cabinets is used to monitor aircraft deceleration, pitch and ground speed
3	Proximity sensing system (air/ground sensor)	Main landing gear MLG car inclination (on the ground)
4	Brake system control unit	Normal power, battery powered
5	Landing system	Controls the landing gear lever position
6	Left and right fuel cut-off switches	Fuel supply or stop fuel supply
7	Electronic and Electrical Center(EEC)	Determine the thruster angle

8.2.3　Electric Break Power Supply System

As shown in Fig. 8.2.12, it is the electric brake power supply, in which there are 4 Electric-Brake Power Supply Units(E-BPSU), which provide driving power for the MLG electric brake. Each E-BPSU obtains 28 V/50 A DC power supply from the transformer rectifier unit of the power supply system in the switchboard P500 or P600.

The E-BPSU provides ± 130 V DC to the corresponding Electric Brake Actuator Controller (EBAC), each E-BPSU has a rated power of 2.5 kW and a peak power of 4 kW. The operating voltage of each electric brake is ±130 V DC. Although it is an intermittent operating system, the power supply is not allowed to be interrupted when the aircraft is taxiing on the ground. Therefore, when the main power supply is 230 V AC, it will generate 28 V DC through the transformer rectifier unit, and then generate ±130 V DC through the DC/DC converter. When the main power supply 235 V AC fails, the voltage of the hot battery busbar will be converted to obtain ±130 V DC.

The left and right landing gear brake controller is supplied by P500 and P600 distribution boards respectively. As shown in Fig. 8.2.12, in the middle of the picture, P500 board on the left and P600 board on the right.

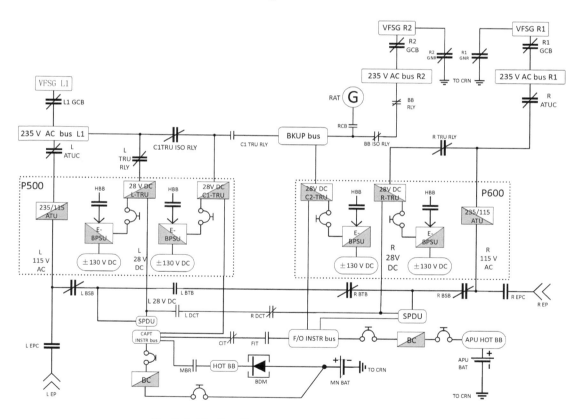

Fig. 8. 2. 12 Landing gear braking power supply system

The E-BSCU obtains 28 V DC from the TRU and supplies it to the brake system control unit. The power distribution of the BSCU is listed in Tab. 8.2.14.

Tab. 8. 2. 14   Power distribution of the BSCU

SN	Name	Distribution object	Distribution board
1	28 V DC C1 – TRU	Left outer brake system control unit	P500
2	28 V DC L – TRU	Left inner brake system control unit	P500
3	28 V DC C2 – TRU	Right outer brake system control unit	P600
4	28 V DC R – TRU	Right inner brake system control unit	P600

If the power supply is invalid from the transformer rectification unit, the hot battery busbar contactor is connected to power the Electric-Brake Power Supply System Unit (E-BPSU).

The HBB connects the main battery to the E-BPSU and the associated BSCU channel. The battery switch on the overhead electrical panel (P5) must be in the ON position. The battery switches will be connected to the E-BPSU respectively, enabling the HBB connector to work.

The towing power switch on the P5 board overhead in the cockpit is connected by the control logic to the E-BPSU to enable the operation of the hot battery bus contact so that the main battery can be used for braking operation during towing.

The left BSCU provides 40 V DC power to the 4 MLG on-axis RDCs on the left main landing gear. The right BSCU provides 40 V DC power to the 4 MLG on-axis RDCs on the right main landing gear and 28 V DC power to the 2 NLG axle RDC. The right BSCU also provides 28 V DC power for the NLG status lights.

8.3   Brake System Operation

Braking is the process of stopping or decelerating the aircraft during driving. MEA braking forms include automatic braking and parking brakes. Before takeoff, if there are conditions that do not meet the takeoff, you can use the brakes to refuse to take off. When the aircraft stops, you must pull up the handbrake to brake, and you need to release the brakes when you takeoff. The size of the braking force can be selected for sub-block braking. In the cockpit, the pilot operates the brake selection switch to achieve the aircraft ground motion control according to the actual needs. The following will introduce the brakes of the aircraft.

8.3.1   Automatic Braking

The B787 has a selector switch, as shown in Fig. 8. 3. 1, it is the automatic brake selector switch, there are 5 positions, namely 1, 2, 3, 4 and the maximum automatic (MAX AUTO), as well as the DISARM position for releasing the brake, the Rejected Takeoff (RTO) position for rejecting takeoff and the closing position. The automatic brake selector switch can be interfaced with the BSCU.

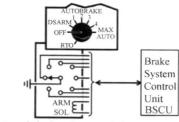
Automatic brake selector switch

Fig. 8. 3. 1   Automatic brake selector switch

The switch is located on the outer channel of the left BSCU to provide a switch input signal. The automatic brake selector switch gears are listed in Tab. 8. 3. 1.

Tab. 8. 3. 1   Automatic brake selector switch gears

SN	Name	SN	Name
1	Rejected takeoff (RTO) position	4	1, 2, 3, 4 four brake gears
2	Closed position OFF	5	Maximum automatic braking
3	Release position DISARM	—	—

When the plane landed should start automatic brake operation, the operating conditions listed in Tab. 8. 3. 2 must be met.

Tab. 8. 3. 2   Start the automatic brake operating conditions upon landing

SN	Operating conditions
1	Selector switch for automatic braking in position 1, 2, 3, 4 or maximum (BSCU applies maximum braking force in maximum automatic position)
2	Aircraft tilt data from the proximity sensing system when the aircraft is in the air
3	When the aircraft is idling, the data of the two thrust rods, among which the data of the thrust rods come from the Thrust Resolver Angle (TRA) data of the electronic and electrical controller and the position information of the thrust rods
4	The brake pedal force is less than 65% of the full value
5	No automatic brakes or skid failures

The BSCU provides automatic control commands for aircraft landing. Tab. 8. 3. 3 lists the operating conditions for landing commands of BSCU.

Tab. 8. 3. 3   Operating conditions for landing commands of BSCU

SN	Operating conditions
1	Automatic braking operation in conjunction with landing
2	At idle, TRA data from electronic and electrical controller for both thrust rods
3	Aircraft tilt data from the proximity sensing system when the aircraft is parked on the ground
4	From the landing gear lever position data, when the gear lever is lowered for more than 20 seconds
5	The average wheel speed is greater than 60 knots(1 kn=1. 852 km/h)

If there is no TRA data or the TRA data is invalid, the left outer BSCU uses the fuel disconnect switch at the cut-off position, that is, when the aircraft is equivalent to the idle speed indication state. Then the off switch data uses the data on the cockpit display and the crew alert function data. As listed in Tab. 8. 3. 4, when one or more of the following situations occur, switch the automatic brake selector switch to the release position to release the automatic brake operation when the plane landed.

Tab. 8. 3. 4　Automatic braking operation when the aircraft is released from landing

SN	Operating conditions
1	The brake pedal is in place
2	Speed brake lever moved to loading position (speed brake lever position in Flight Control Electronics (FCE) cabinet)
3	One or two thrust levers leave the idle position according to the inference resolver TRA data of the EEC
4	Malfunction of automatic braking, anti-skid or braking system

When the aircraft is under certain conditions, it can refuse to takeoff. Tab. 8. 3. 5 lists the situations of refusal to takeoff.

Tab. 8. 3. 5　Conditions for aircraft refusal to takeoff

SN	Operating conditions
1	Automatic brake selector switch in RTO position (BSCU uses maximum braking force in this position)
2	Level data from the aircraft on the tarmac from the approximation sensing system
3	No automatic brake failure

The left outer BSCU gives the auto-brake command to Rejected Takeoff (RTO), as listed in Tab. 8. 3. 6, it is a necessary condition for operation.

Tab. 8. 3. 6　Operating conditions for RTO auto-brake command

SN	Operating conditions
1	The wheel has a possible state of explosion
2	Rolls at 85 knots on takeoff
3	2 thrust levers not at idle, no auto brake or skid failure
4	2 idle thrust rods (TRA data from EEC) with an average wheel speed of over 85 knots

8.3.2　Nose Landing Gear Wheel Steering Brakes

As shown in Fig. 8. 3. 2, it is the nose wheel of the NLG rotates the brake, Fig. 8. 3. 2(a) shows the brake picture, and Fig. 8. 3. 2(b) shows the position of the brake on the aircraft. There are two front wheels on the NLG touching the rotating brake, and turning the clip on the brake makes the front wheels stop moving. When the nose landing gear is retracted into the wheel well, the turning brakes on the nose wheels stop the movement of the nose wheels.

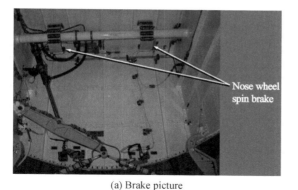 

(a) Brake picture   (b) Position of the brake on the aircraft

Fig. 8. 3. 2   Nose landing gear nose wheel spin brake

8.3.3   Brake Indication

1. The Outline of the Landing Gear

As shown in Fig. 8. 3. 3, the outline of the landing gear is displayed, which shows the temperature data of the brake, the tire pressure data, the fault information of the brake anti-skid, etc. B787 has a total of 10 landing gears. There is a selection menu at the top of the page for calling up to check the status of the landing gear.

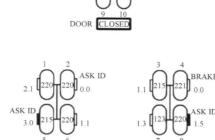

Fig. 8. 3. 3   Landing gear synoptic

The landing gear simultaneously displays the tire pressure for each wheel and the brake temperature indication for the main landing gear wheels. The tire pressure indicator turns amber when the conditions listed in Tab. 8. 3. 7 occur.

Tab. 8. 3. 7   Tire pressure warning display (amber)

SN	Warning condition	Color
1	Tire too low (less than 100 psi)	Amber
2	Tire pressure 25% less than another tire pressure on the Main Landing Gear (MLG) axle	Amber
3	12% less tire pressure than other tires on the Nose Landing Gear (NLG)	Amber
4	18% less pressure than all tires on the same MLG	Amber

The brake temperature of the main wheel is not indicated by the specific temperature, but is indicated by the standardized data display, i. e. 0. 0—9. 9. The brake temperature of the main wheel is represented by a decimal point with a significant number, the normal brake temperature is displayed as a white box, and the hottest MLG brake temperature is displayed in white as a whole. The brake overheat indicator light is displayed in amber, if the brake data is invalid, the displayed symbol is normal, but the data is empty. The brake temperature warning information is displayed on the EICAS, and the brake temperature indication is greater than 5. 0. This value remains unchanged until the brake temperature indication is less than 3. 0, which is the brake temperature indication information. Relevant brake information can be displayed above the brake temperature indicator. These temperature information can be displayed next to the main wheel outline.

When braking, the BSCU detects a brake failure message. When skidding, the BSCU detects a skid failure, or receives a communication failure from the MLG Axle RDC.

2. Landing Gear Brake Maintenance Page

As shown in Fig. 8. 3. 4, it is the maintenance page of the landing gear brake, which is used for maintenance by the crew. The maintenance page shows the brake data listed in Tab. 8. 3. 8.

Tab. 8. 3. 8   Brake data displayed on the maintenance page

SN	Display content
1	Brake pedal position, captain position and flight control position signals
2	Parking handbrake status, if the relevant data is invalid, it will be blank
3	There are tire pressure values in each nose wheel profile
4	Brake temperature, tire pressure, and brake life are displayed in each main wheel outline symbol, and the force of the electric brake actuator is displayed near the main wheel outline symbol

For convenience, the messages listed in Tab. 8. 3. 9 can be displayed next to the wheel profile.

Tab. 8. 3. 9   Shows the message next to the wheel profile

SN	Display content
1	Displayed for main wheels if there is an electric brake failure or electric brake retraction failure

SN	Display content
2	The brake skid control of the main wheel is malfunctioning
3	Loss of functionality related to RDC on the main landing gear axle
4	Main wheel brake ground test deactivated
5	The main wheel brake has an electric brake actuator failure

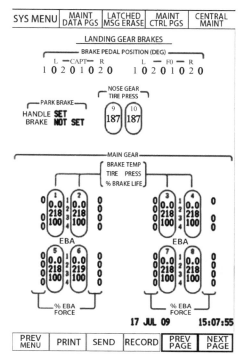

Fig. 8.3.4　Maintenance page of the landing gear brake

Related event messages can be displayed at the bottom of the page.

3. Nose Landing Gear Status Lights

As shown in Fig. 8.3.5, it is the nose landing gear status light indicator, which is red

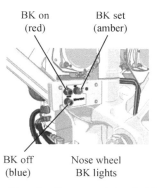

Fig. 8.3.5　Nose landing gear status light indicator

when the brakes are on, amber when the brakes are set, and blue when the brakes are off.

8.3.4 Parking Brake

When the aircraft needs to stop, the brake system must be used and ensure that it does not move on the ground. This involves the setting of the stop brakes, towing the aircraft, etc.

1. Parking Brake Settings

The process of stopping the aircraft requires the use of the parking brake lever. As shown in Fig. 8.3.6, it is the schematic diagram of the parking brake lever of the B787. There are 2 positions for the parking brake of the aircraft. The first position is in the lower position, and the aircraft can move at this time. The second position is in the upper position when the handbrake is pulled up, and the aircraft cannot move at this time.

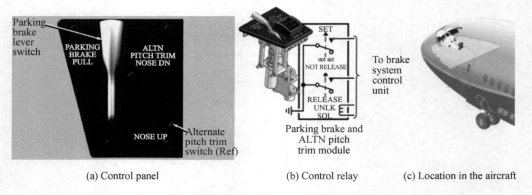

(a) Control panel   (b) Control relay   (c) Location in the aircraft

Fig. 8.3.6 Schematic diagram of parking brake lever (B787)

If the brake pedal is not pressed, the brake lever can only be in the down position, at this time, the solenoid valve prevents the lever from moving up. When the captain or co-pilot depresses the brake pedal and the handbrake is pulled up, the brake pedal triggers the BSCU to send a signal to disconnect the operating power of the brake lever solenoid valve, so that the parking brake lever is locked in the upward position.

Use the parking brake lever to hold the brake to stop, there are 2 internal switches to control the sequence of the parking brake, as shown in Fig. 8.3.6. Fig. 8.3.6(a) shows the control panel, which has a "Parking brake lever" and a double-pole double-throw switch. Fig. 8.3.6 (b) shows the hand-pull Brake Electromagnetic Control Relay(BECR). Fig. 8.3.6 (c) shows its position on the aircraft. When the parking brake is in the position shown in the figure (down position), the relay contact is in the position shown in Fig. 8.3.6(b). That is, the upper contact of the relay is in the "NOT SET (no setting)" position, and the lower contact is in the "RELEASE (release)" position. When the parking brake is pulled up, the upper contact of the relay is in "SET (set)", the lower contact is in "NOT RELEASE (not released)", each contact signal will be sent to the BSCU.

The relay coil in Fig. 8. 3. 6(b) allows free movement of the brake lever when energized, and leaves the brake lever in the hold position (up or down) when de-energized. The control signal of the relay comes from the BSCU control, there is no mechanical connection between the parking brake lever switch and the pedal.

When stopping braking, the braking force needs to be adjusted, and the braking force should be adjusted to more than 25% of the maximum value. If the EEC sends the TRA data to the thrust lever in the non-idle position, the braking force will increase to 100%. The 2nd brake position switch is closed, releasing the brake pedal causes the BSCU to de-energize the brake lever solenoid valve.

There is no mechanical connection between the pedals and the parking brake lever switch, when the brake pedals are released, the spring reaction force returns the brake pedals to the up position.

After the handbrake is pulled, the ±130 V DC power supply is connected for 1 minute, and the EBAC can adjust the EBA and then send a power request to the BSCU before the EBA adjustment can be performed.

When the parking brake is set, the 60-minute timer in the BSCU starts and the BSCU remains in the parking adjustment mode, which allows the EBAC to adjust the EBA as the brake temperature changes. Allows the braking system to be powered by the main battery if the battery switch is closed before the end of 60 minutes.

When the indicator light is powered on, the status light on the top of the nose gear displays the brake indication (as shown in Fig. 8. 3. 5)

① Amber indicator light: Set the parking brake.

② Red indicator light: Brake status.

③ Blue indicator light: No brake state.

④ If the red light is on, the blue light will come on when the braking force drops below 2% of the maximum value. If the blue light is on, the red light will turn off when the braking force increases above 25% of the maximum value.

Therefore, the braking operation sequence for aircraft shutdown is summarized as follows:

① The parking handbrake lever moves up, because there is a switch inside, use this switch to send a switch trigger signal to the brake power system controller.

② Send the brake pedal signal to the BSCU, the internal solenoid valve is energized, so that the parking brake lever can move to the specified position.

③ The trigger signal from the second switch inside makes the BSCU set the parking brake operation.

④ Releasing the brake pedal causes the BSCU to place the solenoid controlling the parking brake lever switch in the hold position.

When the landing gear lever is moved from the up position to the down DN position, the BSCU will provide enabled power to the EBAC. After power up, the EBAC will perform a

Brake Operation Test (BOT).

This sets the brake operation interval. Sixty seconds after the landing gear lever is moved to the UP position, the EBAC power is disconnected.

This will result in the disconnection of the EBAC during flight. When the landing gear lever is moved to the down position, the BSCU sets the power to the brake actuator control mode. If the radio altimeter data shows 500 feet (152 meters) or more, the EBAC will perform a brake operation test, setting the EBA to the pre-landing operating distance.

2. Towing

Towing the power switch (on the P5 board of the cockpit overhead panel), turn on the switch main battery to make the BSCU and E-BPSU work. There is a display of the battery capacity on the battery indicator, and the towing operation is performed according to the battery capacity.

As shown in Fig. 8. 3. 7, it is the battery test switch panel; there is a towing power switch (TOWING POWER), a battery test (TEST) button, and three battery test indicator lights on the panel.

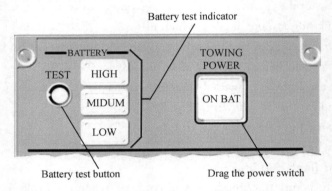

Fig. 8. 3. 7　Battery test switch panel

The towing power switch connects the electric brake power and other necessary systems to the main battery for towing. When the towing power switch is turned on, press the battery test button, according to the battery capacity, the battery indicates the capacity and the towing operation mode in three states: high, medium and low. The battery indicator shows the charging voltage of the main battery:

① High: The towing operation time is at least 60 minutes.

② Medium: The towing operation time is at least 30 minutes.

③ Low: The towing operation time is at least 15 minutes.

As shown in Fig. 8. 3. 7, the brake can reduce the minimum time of the high, medium and low voltage states during towing. If the battery voltage changes from high to medium during towing, the towing must be completed within 30 minutes of the medium duration. If the battery voltage changes from the mid-range to low-range during towing, the aircraft must be towed to a parking position where an external power supply can be connected or the

APU can be activated. The battery power of the brake pedal operation is the same as in other aircraft, and the brakes of the front and rear wheel pairs during towing are the same as when the aircraft rolls.

8.4 Nose Wheel Steering System

8.4.1 Introduction

In addition to supporting the weight of the aircraft, the Nose Wheel Steering System (NWSS) is also responsible for functions such as nose steering. Fig. 8.4.1 shows the outline of the NWSS.

Fig. 8.4.1   Outline of the NWSS(B787)

Fig. 8.4.2 shows the block diagram of the NWSS. The picture shows the nose wheel, reversing valve, actuator, valve control module, Remote Electronic Unit (REU), flight control electronic equipment, Tiller Module(TM) and RDC module, etc. Tab. 8.4.1 lists the basic composition of the NWSS.

Tab. 8.4.1   Basic composition of the NWSS

SN	Name	Number	SN	Name	Number
1	Joystick tillers (captain and co-pilot)	2	5	Brake actuator	2
2	Remote electronic unit module	2	6	Rotary valve	2
3	Position sensor module	2	7	Steering wheel	1
4	Valve assembly	1	—	—	—

The NWSS has interfaces with the central hydraulic system, the FCE and the CCS connected via the RDC. There are also interfaces for the Landing Gear Actuators and Nose Wheel Steering (LGA/NWS) control functions in the CCS. Fig. 8.4.3 shows a schematic diagram of the NWS control. The steering input signal is transmitted from the joystick and the flight control electronic system FCE to the rudder pedal position signal to the landing gear actuator/near wheel steering system. The rudder pedal disconnects the switch on the

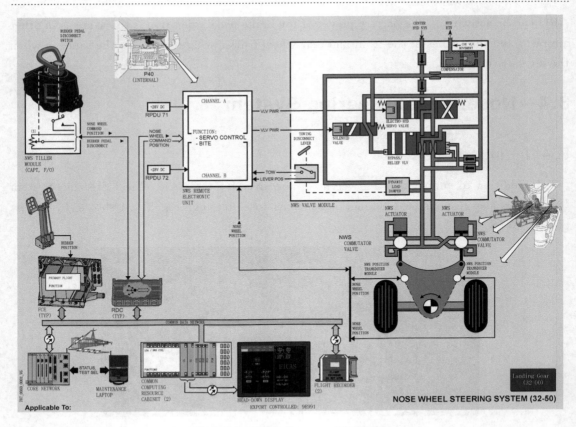

Fig. 8. 4. 2　Block diagram of the NWSS

joystick and disconnects the control input signal to the Flight Control Electronics(FCE) to the landing gear actuator/NWSS. The LGA/NWS calculates steering commands and transmits these commands to the CDN. The NWSS REU uses the commands to control the NWS valve control module.

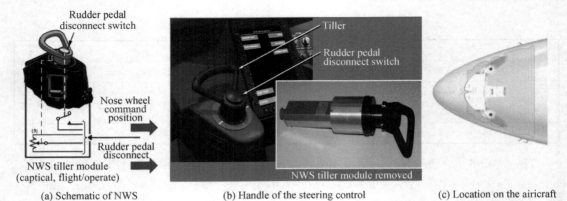

(a) Schematic of NWS　　　(b) Handle of the steering control　　　(c) Location on the airicraft

Fig. 8. 4. 3　Schematic diagram of the NWS control

　The valve control module of the NWS system controls the 2 actuators by using the central hydraulic system, and the actuators turn the NWS rim of the nose wheel. The position sensor module of the NWSS provides the position signal input of the nose wheel rotation to the NWS REU, and the REU performs servo control. It can be seen from Fig. 8. 4. 2 that the NWS REU obtains 28 V DC from the remote control RPDU. There are remote power distribution units RPDU71 (left DC busbar) and RPDU72 (right DC busbar) for power supply routes.

① When the speed of the aircraft is less than 30 knots, the maximum angle that the nose wheel can be rotated by turning the joystick is ±70°.

② When the speed of the aircraft is between knots, turn the joystick to decrease the speed of the aircraft in the LGA/NWS control function.

③ When the speed of the aircraft is greater than 100 knots, turning the joystick can make the rotation angle of the nose wheel to be ±8°.

8.4.2 NWS Tiller Module

Turning the internal potentiometer for each steering joystick according to Fig. 8. 4. 3(a) gives an output related to the joystick position. The output of each sensor is connected to the CDN through different Remote Data Concentrates (RDC). Potentiometer in the captain's joystick provides inputs for RDC 1, RDC 3, and RDC 11. Potentiometer in the co-pilot joystick provides inputs for RDC 2, RDC 4, and RDC 12. RDC changes the analog input to the digital signal and transmits the digital signal on the CDN. Rudder position data is transmitted to the CDN via the Flight Control Electronics (FCE). The FCE also provides inputs for the NWS auto land and thrust asymmetry control.

The control devices are installed on the joystick module, which are respectively the steering joystick and the rudder pedal disconnect switch, so the joystick module can be manually controlled in the cockpit to steer the nose wheel. If both joysticks are turned in the same direction, the value corresponding to the position of the joystick sensor is added together; if the joystick is turned in the opposite direction, the value corresponding to the position of the separate joystick is subtracted together.

During the test and maintenance, the rudder pedal disconnect switch on each NWS sub-rudder module will delete the rudder pedal position input in the calculation of the steering command, and allow the movement of the rudder to be tested without the movement of the nose wheel.

8.4.3 Nose Wheel Steering Remote Control Electronic Unit

The NWS REU provides servo control of the nose wheel position. REU has two internal signal channels, during normal operation, one channel is used for control and the other channel is reserved.

The NWS REU receives control commands from the LGA/NWS and compares this control command with the nose wheel position data from the position sensor module. The difference between the commanded position and the nose wheel position controls how far the

NWS REU moves the control valve in the NWS valve module.

The NWS REU sends the nose wheel position command data into 2 NWS REU channels, each channel has command sub-channel and monitor sub-channel.

The data of the CDN is transmitted to the nose wheel steering controller through the RDC. The corresponding relationship between the command channel and the RDC is listed in Tab. 8. 4. 2.

Tab. 8. 4. 2　Corresponding relationship between the command channel of the nose wheel steering system and the RDC

SN	Name	Corresponding RDC
1	Channel A, command sub-channel	RDC 1
2	Channel A, monitor sub-channel	RDC 3
3	Channel B, instruction sub-channel	RDC 2
4	Channel B, monitor sub-channel	RDC 4

It is worth noting that only 1 NWS REU channel can be in control at a time. The other channel must be in standby. When the Nose Landing Gear (NLG) retracts to the up and locked position, the channel in control momentarily changes to standby. This lets the other channel then change from standby to in control if it can. If the channel cannot make this change, the first channel changes back to in control again. As shown in Fig. 8. 4. 4, it is the NWS remote electronic assembly diagram.

NWS REU

Fig. 8. 4. 4　NWS remote electronic assembly diagram

8.4.4　Nose Wheel Steering Valve Module

The NWS valve module is mainly composed of nose wheel steering valve module, steering actuator, AC valve, position sensor module, steering operation and tow bar, etc.

1. NWS Valve Module

The NWS valve module controls the central hydraulic system of the NWS actuator, as listed in Tab. 8. 4. 3, it is the internal part of the NWSS valve module.

Tab. 8. 4. 3　Internal part of the NWSS valve module

SN	Name	SN	Name
1	Electro-Hydraulic Servo Valve(EHSV)	6	Reduced oscillation of the front wing wheel
2	Solenoid operated test valve	7	Compensator
3	Bypass relief/safety valve	8	Internal pressure maintainer
4	Actuator fluid allows controller	9	Drag switch
5	Dynamic load damper	—	—

Fig. 8. 4. 5 shows the schematic diagram of the NWS actuator valve module. The NWS REU controls the operation of the Electro-Hydraulic Servo Valve (EHSV) and solenoid valve operation test valve SOPV, drags the disconnect lever to set the Dynamic Load Damper (DLD), and allows the nose wheels to turn freely during towing. The tow switch provides NWS REU indication of the tow disconnect lever position.

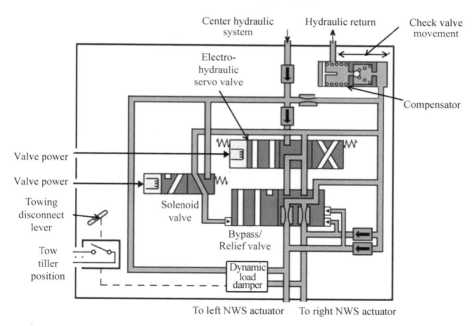

Fig. 8. 4. 5　Schematic diagram of the NWS actuator valve module

Solenoid Operated Pilot Valve(SOPV) and EHSV have two solenoid coils each. Each SOPV solenoid valve coil is connected to one NWS REU channel. The command sub-channel is connected to the solenoid valve power supply, and the monitor sub-channel is connected to the solenoid valve circuit. If a fault occurs, the command sub-channel or the monitor sub-channel can be connected to the solenoid valve coil circuit.

Each EHSV solenoid valve is connected to one command sub-channel in each NWS REU channel. The other internal components in the NWS valve module are hydro-mechanical components, the bypass relief/safety valve, which allows fluid to flow into the actuator;

DLD (steering wheel damping device) to dampen nose wheel oscillations. The function of the compensator is to maintain the internal pressure at the DLD if the central hydraulic system loses pressure.

The NWS valve module controls the central hydraulic system pressure to the 2 NWS actuators that turn the nose wheels. The NWS actuator body is mounted between the steering plates on the outer strut cylinder, the actuator rod is connected to the steering wheel, and the actuator can push or pull to move the steering wheel so that the inner cylinder and the nose wheel are connected by torque.

2. NWS Actuator

The NWS actuator provides the force when moving the steering nose wheel. As shown in Fig. 8.4.6, it is the nose wheel steering actuator and valve module. There is one actuator on the left and right in the picture, and the NWS valve module controls the central hydraulic pressure that operates the 2 NWS actuators. The NWS actuator turns the nose wheel through the NWS and torque link.

Fig. 8. 4. 6   NWS actuator and valve module

3. NWS Commutator Valve

The NWS divert valve controls the NWS actuator port to obtain hydraulic pressure relative to the position of the nose wheels. The NWS actuator rotates relative to the NWS divert valve. When the angle of rotation of the nose wheel is small, the two actuators work in coordination, one of which is in the state of pushing, and the other must be in the state of pulling. When the nose wheel turns, the actuator in the pull state will go to the over-center position. At this angle, the associated NWS divert valve changes pressure to the end of the actuator, causing it to push.

As shown in Fig. 8. 4. 7, it is the NWSS steering valve, which shows the reversing valve, valve module and actuator, etc. When the steering angle is less than 27°, one actuator is pushed and the other is pulled. When the steering wheel is turned, the NWS actuator rotates relative to the reversing valve.

When the steering ring turns, the NWS actuator turns relative to the divert valve. At about 27°, the actuator with rod end pressure is perpendicular to the strut, and the divert

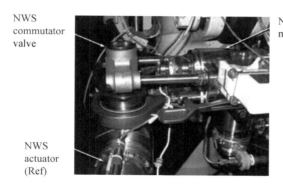

Fig. 8. 4. 7 NWSS reversing valve

valve passes through the changing pressure port to provide head end pressure to the actuator. Both actuators can push and pull together to move the nose wheel to a maximum position of ±70°.

4. NWS Position Sensor Module

The NWS position sensor module provides the feedback signal of the nose wheel position to the NWS remote electronic module. As shown in Fig. 8. 4. 8, it is the NWS position sensor module.

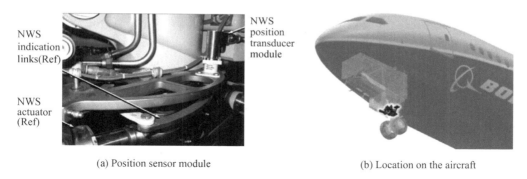

(a) Position sensor module   (b) Location on the aircraft

Fig. 8. 4. 8 NWS position sensor module

Each NWS position sensor module contains two Rotary Variable Differential Transformers (RVDT). The RVDTs provide analog position data to their respective NWS REU channels. The NWS REU channel provides the RVDT excitation. Each module is located on a different mounting plate. The NWS indication link connects the shaft of the sensor module to the steering wheel that turns the nose wheels.

5. Steering Operation

Steering operation has a nose wheel steering joystick and rudder pedals to provide manual steering input. A disengage switch on the rudder pedal allows for rudder movement tests without nose wheel movement. Steering data is shown at the bottom of the Landing Gear Indication and Steering (LGIS) maintenance page. The steering system is tested for proper operation after the landing gear is down.

(1) NWS Control Input

There are two NWS joystick modules for the commander and the first officer. The joystick and rudder pedals provide position input signals for steering control. Each joystick module has 3 potentiometers that turn according to the handle position. Each potentiometer provides an analog input signal to a different Remote Data Concentrate (RDC).

Potentiometer in the captain's joystick provides inputs for RDC 1, 3, and 11, and potentiometer in the copilot's joystick provides inputs for RDC 2, 4, and 12. The RDC converts the analog input to a digital signal and places the data on the CDN.

Rudder position data is sent to the CDN via the Flight Control Electronics (FCE). The FCE also provides input for autoland and thrust asymmetry control.

(2) NWS Position Command

The LGA/NWS control functions in the CCS use this data to calculate the nose wheel position commands. If both joysticks are moved at the same time, the LGA/NWS control function will calculate a nose wheel position command relative to the sum of the joystick positions. Each NWS joystick module has a rudder pedal disconnect switch that disconnects the FCE input signal from the calculated nose wheel position command at aircraft speeds less than 40 knots. This allows the rudder to operate without input to the NWS at low speeds or when the aircraft is parked.

At different speeds of ground planes, the range of rotation of the nose wheels is different, the faster the speed, the smaller the range of rotation. As listed in Tab. 8. 4. 4, when the aircraft is on the ground, the steering angle of the nose wheel of the aircraft is controlled by the joystick and the pedal.

Tab. 8. 4. 4   NWS angle and aircraft speed range

SN	Aircraft ground speed range/knot	Maximum angle of nose wheel/(°)	Rotation input control
1	≤30	±70	Joystick manipulation
2	≥100	±8	Joystick manipulation
3	—	±8	Pedal input
4	30—100	±70	Joystick manipulation
5	—	±70	Joystick and pedal combination

LGA/NWS control function sends nose wheel position commands to CDN.

(3) NWS Action

The NWS REU provides servo control to turn the nose wheel to the commanded position. The nose wheel position command data goes into two NWS REU channels A and B. Each channel has an instruction sub-channel and a monitoring sub-channel. Data from CDN is transmitted to NWS REU through RDC. The RDC distribution of channel functions is listed in Tab. 8. 4. 5.

Tab. 8. 4. 5   RDC distribution of channel functions

SN	Channel name	Sub-channel	Corresponding RDC
1	Channel A	Command channel	RDC 1
2	Channel A	Monitor sub-channel	RDC 3
3	Channel B	Command channel	RDC 2
4	Channel B	Monitor sub-channel	RDC 4

Left and right NWS position sensor modules monitor the nose wheel position, and each module has two RVDTs.

The RVDT in the NWS position sensor module is used as the excitation signal in the NWS REU, and each NWS position sensor module has a calibration procedure. As listed in Tab. 8. 4. 6, it is the function of the RVDT in the position sensor.

Tab. 8. 4. 6   Functions of RVDT in the position sensor

SN	NWS position sensor	Corresponding RVDT	Features
1	Right	RVDT 1	B-channel command sub-channel
2	Right	RVDT 2	Channel A monitor sub-channel
3	Left	RVDT 1	Channel A command sub-channel
4	Left	RVDT 2	Channel B monitor sub-channel

Solenoid Operated Pilot Valve(SOPV) and EHSV have two solenoid coils each. Each SOPV solenoid valve coil is connected to one NWS REU channel. The command sub-channel is connected to the solenoid valve power supply, and the monitor sub-channel is connected to the solenoid valve circuit. If a fault occurs, the command sub-channel or the monitor sub-channel can be connected to the solenoid valve coil circuit.

Each EHSV solenoid valve is connected to one command sub-channel in each NWS REU channel. The other internal components in the NWS valve module are hydro-mechanical components, the bypass relief/safety valve, which allows fluid to flow into the actuator. Dynamic Load Damper (DLD) (steering wheel damping device) is used to dampen nose wheel oscillations; the compensator function is to keep internal pressure at 240—270 psi. The DLD compensation if the central hydraulic system loses pressure.

The NWS valve module controls the central hydraulic system pressure to the two NWS actuators that turn the nose wheels, the body of which is mounted between the steering plates on the outer strut cylinders. The actuator rod is attached to the steering wheel and can push or pull and move the steering wheel so that the inner cylinder and the nose wheel are connected by torque.

When the steering angle of the nose wheel is less than 27°, one actuator is pushed by front end pressure and the other is pulled by rod end pressure.

When the steering wheel is turned, the NWS actuator rotates relative to the diverter valve. At approximately 27°, the actuator with rod end pressure is perpendicular to the strut, causing the diverter valve to change the pressure port to provide front end pressure to the actuator. In this way, the two actuators push and pull together to achieve the maximum rotation range±70° of the nose wheel.

(4) The Nose Landing Gear is Lowered

When the NLG is down indicating the proximity sensor, indicating that the NLG is down, the LGA/NWS control function performs the following two-step test:

① Immediately give a nose wheel steering 3° position command, and open the solenoid operated pilot valve.

② Then close the solenoid operated pilot valve immediately and give a position command to turn the nose wheel −3°.

The first part of the test is to ensure that the nose wheels are moving to the correct angle, and the second part of the test ensures that the NWS REU can close the SOPV to prevent the nose wheels from moving. Failure to test will result in a NWS warning message on the EICAS.

An internal cam aligns the nose wheel to a center position when the internal hydraulic column of the nose wheel strut is extended. The NWS system can move the nose wheel out of the cam-centered position.

(5) Towing Operations

Fig. 8.4.9 shows the position control of the nose wheel landing gear tow rad, the tow disconnect lever on the NWS valve has a position control latch that allows the nose wheel to move freely during towing operations. Fig. 8.4.9(a) shows the tow bar in the normal position, Fig. 8.4.9(b) shows the tow rod in the tow position, the latch in the tow bar holds the tow disconnect lever in the tow position.

A dynamic load damper is mechanically set to move hydraulic fluid between the NWS actuators. An internal switch in the NWS valve module causes the two NWS REU channels to stand by. This will de-energize the solenoid operated pilot valve and set the bypass safety valve/safety valve to the bypass position.

As shown in Fig. 8.4.10, the nose landing gear torque connector is disconnected, the nose wheel can be turned to a maximum of 70° for towing, and it can be quickly disconnected at the apex of the torque connection to allow the nose wheel to rotate 360°. The cable on the torque connection must be disconnected before releasing the quick disconnect.

Tow lever in the normal position

Pin holds tow lever in the towing position

Tow lever
in the towing
position

(a) Normal position

(b) Tow position

Fig. 8. 4. 9   Position control of the nose wheel landing gear tow rod

Top torque link

Bottom torque link

(a) Nose landing gear

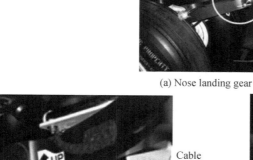

Cable
stowage
connection

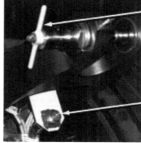

Quick
disconnect
handle

Cable
normal
connection

(b) Top torque connection

(c) Bottom torque connection

Fig. 8. 4. 10   Nose Landing gear torque connector disconnect

8.5   Ground Maintenance

Testing of the brake control system is for the inner and outer channels of each Brake System Control Unit (BSCU).

8.5.1 Ground Maintenance Platform

Fig. 8.5.1 shows the schematic diagram of the ground maintenance platform. In the figure, a laptop is used for information query and operation. Combine the data of the CDN with the laptop maintenance platform through the core network to conduct ground maintenance tests on the entire brake system.

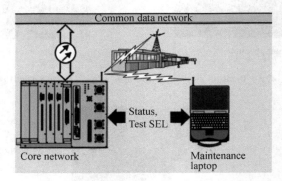

Fig. 8.5.1 Ground maintenance platform

The ground test (GROUND TESTS) is displayed on the test software to open the menu to display the drop-down menu on the right side of Fig. 8.5.2. Select the test type through the cursor, system test (SYSTEM TEST), the content of SN bit 32 displayed in the menu is the test of the braking system of the landing gear.

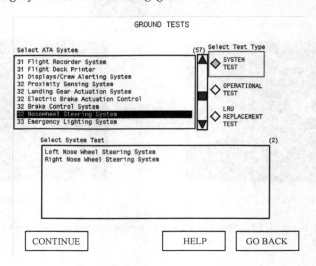

Fig. 8.5.2 Ground test page of brake control system

There are mainly proximity sensing system, landing gear actuator system, brake control system and nose wheel steering system menu. Under each menu, there is a menu of the next layer. Take 32 Brake Control System as an example, the drop-down menu has options for system tests as follows:

① Left Nose Wheel Steering System.

② Right Nose Wheel Steering System.

8.5.2　Ground Test Menu

As shown in Fig. 8.5.3, it is the electric brake actuator ground test menu. There are 8 drop-down menus under ground test and electric brake actuation control, which are:

① Wheel 1 Brake Actuation System.

② Wheel 2 Brake Actuation System.

③ Wheel 3 Brake Actuation System.

④ Wheel 4 Brake Actuation System.

⑤ Wheel 5 Brake Actuation System.

⑥ Wheel 6 Brake Actuation System.

⑦ Wheel 7 Brake Actuation System.

⑧ Wheel 8 Brake Actuation System.

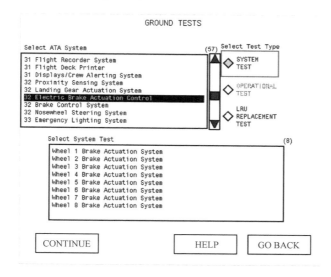

Fig. 8.5.3　Ground test menu controlled by the electric brake actuator

8.5.3　Testing of Special Functions

In addition, there are some special function ground test menus, as shown in Fig. 8.5.4. Select Electric Brake Actuation Control, the function selection test of the drop-down menu will appear, mainly as follows:

① Wheel 1 Brake/EBA Deactivation.

② Wheel 1 Full Break Retract.

③ Wheel 2 Brake/EBA Deactivation.

④ Wheel 2 Full Break Retract.

⑤ Wheel 3 Brake/EBA Deactivation.

⑥ Wheel 3 Full Break Retract.

⑦ Wheel 4 Brake/EBA Deactivation.

⑧ Wheel 4 Full Break Retract.

⑨ Wheel 5 Brake/EBA Deactivation.

There are a total of 8 aircraft wheels. The test page of the following content can be displayed by the scroll bar on the right side of the figure. After selection, the ground test can be performed.

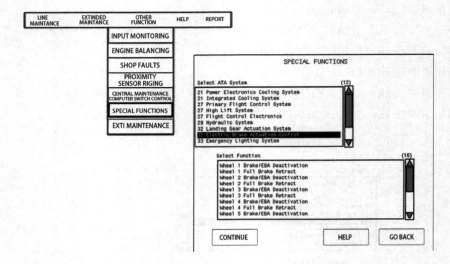

Fig. 8. 5. 4   Special function ground test menu

8.6   Summary

This chapter takes the MEA braking system as the object to introduce, mainly introduces the landing gear braking system of B787. The landing gear braking system is mainly composed of braking system control power supply, electric braking system actuation controller and various sensors. It also introduces the remote data communication network of the brake system, the operation of the brake system, the nose wheel steering system and the ground maintenance, etc.

The design method of the braking system of MEA utilizes new technologies, such as sensor information technology, remote data communication network technology and power electronics technology, which eliminates the hydraulic system and greatly improves the volume, weight, efficiency and reliability.

8.7 Exercises

1. What are the functions of the landing gear?
2. What are the operations of the brake system?
3. What are the main components of the landing gear brake system?
4. What is the main role of the front-wheel steering system?
5. What method may be used to repair the whole brake system on the ground?

Chapter 9 Thermal Management and Energy Optimization of MEA

Compared with conventional aircraft, MEA includes higher power levels and more power-using devices. As electrification increases, the interactions between power sources and power-using devices continue to increase, especially in terms of electrical energy transfer, heat generation and heat dissipation.

Since each electrical component generates heat when it operates and performs loss calculation by the method of power consumption generating heat. The MEA that couples electrical energy, thermal energy and engine is analyzed based on the integrated layout of the electrical system and dynamic changes in energy generation, as well as the steady-state and transient operating conditions of the motor, power converter, storage battery, transformer and each electrical load. This is the goal of energy management and thermal management of an MEA.

The energy of MEA mainly comes from aviation fuel. The storage capacity of the aircraft's chemical battery and compressed air bottle is very limited, so the efficient use of aviation fuel is the key to aircraft and engine energy management. An aircraft that no longer extracts the compressed air from the engine becomes a more desirable MEA. The energy management system actually not only includes the management of electrical energy, but also includes the efficient energy conversion of the aircraft power generation system. MEA uses inverter alternators or high-voltage DC generators, and the energy conversion efficiency is already much higher than the efficiency of traditional aircraft electrical energy generation methods.

9.1 Characteristics of MEA Power Equipment

The B787, A380 and F35 are representatives of MEA/AEA, and the main features of their electrical equipment compared to conventional aircraft are:

① High power of electricity-using equipment.

② Most of the power-using equipment is pulse power equipment, and the ratio of instantaneous power to average power is in the range of 3 to 10.

③ High-power actuators not only absorb electrical energy, but also generate high-power transient braking return energy.

④ For fighter aircraft, laser weapons are inefficient and generate not only large pure power when excited, but also a large amount of electrical energy loss and heat, which will damage other equipment inside the aircraft and the laser weapon itself, if the heat is not handled properly.

Therefore, an MEA must have not only an electrical energy management system, but also a thermal management system.

The B777 aircraft, born in the 1990s, was the first modern large transport aircraft to adopt an automatic electrical load management system. The adoption of automatic load management has reduced the burden on pilots and improved economic efficiency, improved the reliability and maintainability of the aircraft's electrical system, and reduced the aircraft's whole life cycle costs.

The contactors in B777 still use electromagnetic controller, while B787 uses solid-state power controllers, which further improves the reliability and maintainability of the load management system, simplifies the system structure, however, this kind of load management system belongs to steady-state load management, which is designed to achieve the balance between output power and load demand power, and prevent the aircraft power supply system from being overloaded for a long time.

In the B777 and B787, in addition to the load management system, there is also the automatic management of the busbar and power supply, realized with the help of the Busbar Power Management Unit (BPCU), in order to cope with different flight conditions and different failures of the power supply, and to ensure a continuous and reliable power supply to the aircraft, with the power supply and busbar management interconnected with the load management system via a bus.

In terms of aircraft power and load management systems, the B787 and B777 management systems are only steady-state power management systems. The next generation of advanced aircraft power management must also include solutions for transient power management to meet the needs of high-power electronics, electric actuators, and direct energy weapons on military aircraft.

In the aircraft DC grid, there are quite a few energy storage capacitors, filter capacitors at the output of the brushless DC generator, capacitors at the input of the power electronic converter of the speed control motor, filter capacitors at the input and output of the switching power supply, and the storage battery connected to the grid as a large capacitor, and the flat two-wire feeder line also has distributed capacitance. Let the total capacitance of the aircraft grid is C, the grid voltage is U, then the capacitive energy storage in the network is $E = CU^2/2$. When the grid is suddenly connected to a high-power load, the load absorbs the energy of the capacitor, and the grid voltage must have a transient drop; conversely, when the sudden unloading of the load, the voltage must have a transient increase. Similarly, the high-power actuator starting acceleration or braking deceleration, will also be instantaneous absorption or return energy, so that the voltage of the electrical network sudden changes, and even voltage oscillations, thus affecting the operation of all equipment in the electrical network, especially so that it will affect the work of electronic equipment and computers, because the integrated circuit is very sensitive to voltage. In order to make the amount of transient changes in the electric network voltage limited to the specified range, there must be an MEA power management system. The system should not only be able to manage the electrical energy of the aircraft power supply and power-using equipment, reasonably manage the power-using equipment according to the needs of flight and the status of the power supply and busbar, so as to give full play to the potential capacity of the power supply and power-using equipment and save electrical energy, but also should be able to monitor the load and the transient power changes of the power supply to prevent excessive voltage fluctuations of the power grid.

　　Since the transient response times of avionics, actuators and direct energy weapons are only milliseconds to tens of milliseconds, the energy management systems of next-generation MEA must also have fast detection sensors, high-speed processors and efficient software.

9.2　Generation and Storage of Electrical Energy

　　In order to meet the power demand of MEA, there are three methods of electricity generation, namely conventional power generation, high overload capacity power generation and superconducting power generation.

9.2.1　Conventional Power Generation Method

1. Generator Overload Condition

　　Conventional power generation method, increase the electrical energy storage equipment to meet the peak power requirements. The alternators current of MEA are three-stage brushless synchronous generators. Synchronous generators in constant-frequency AC power supply are generally required to be able to overload 100%. Generators in variable frequency AC power supply do not generally require 100% overload due to their larger rated capacity.

　　The generator in the 270 V high-voltage DC power supply consists of a combination of a synchronous generator with a diode rectifier filter circuit, a switched reluctance generator and a biconvex pole generator, all three of which also have less than 100% overload capacity.

　　The permanent magnet generator also has an overload capacity of only 100%, and the F-35 aircraft's switched reluctance S/G rated at 250 kW generates 330 kW with an overload multiplier of only 32%.

　　The overload capacity of both electrically excited synchronous generators and reluctance generators is small. In order to meet the peak power requirements of power-using equipment, the following methods are usually used: firstly, increase the rated capacity of the generator, thus increasing the volume and weight of the motor; secondly, the generator plus electrical energy storage device scheme is used to provide peak power with the help of the storage device.

2. Energy Storage

Lithium-ion battery, super capacitor and flywheel electric generator are typical energy storage devices. Lithium-ion batteries are the highest power density and energy density chemical power sources that can assume instantaneous power and are directly connected to the grid in parallel.

Super-capacitors have high power density but low energy density and can be connected to the aircraft grid through bi-directional converters.

Flywheel electric generator stores and releases mechanical energy with the help of flywheel, and then converts into electrical energy by electric generator, which has high power density. Electric generator has two operating states of electric motor and generator, which realizes the conversion of mechanical energy and electrical energy.

When storing energy, the electric generator works in the electric motor state, and the motor speed is increasing, thus increasing the energy stored in the flywheel. When releasing energy, the flywheel speed decreases and changes from motor state to generator state, converting mechanical energy into electrical energy. The battery, super-capacitor and flywheel motor are bi-directional energy converters, which can absorb electrical energy and release it.

The combined work of conventional generators and energy storage devices can meet the peak power demand of power-using equipment without increasing the rated power of generators, and can also absorb the energy fed back by electric actuators.

A 540 V high-voltage DC PSS ground test platform has been built by Rolls-Royce UK to study the system characteristics with high-power actuators and electrical energy storage devices, such as super-capacitors. The changes in the transient voltage of the test grid without and with energy storage appliances are verified on the experimental platform by testing with and without super-capacitors, in the form of sudden addition and removal of dynamic loads, thus further verifying the smoothing effect of energy storage appliances on the transient voltage changes of the grid.

The platform has two generators: a 70 kW five-phase permanent magnet fault tolerant generator with a 3 000 r/min rotational speed driven by the engine's fan shaft, and a 30 kW switched reluctance S/G driven by the engine's high voltage rotor with a 1 500 r/min rated speed.

The system has three equivalent resistive loads with resistance values varying in the range of 20 to 120 Ω, and four equivalent actuators, each of which rated at 30 kW dynamic load, and a storage device consisting of a super-capacitor.

9.2.2 High Overload Capacity Power Generation Method

The combination of conventional generators and energy storage devices can meet the peak value load requirements, but increases the size and weight of the storage devices. The use of a generator with high overload capacity allows the storage device to be omitted.

The overload capacity of a generator depends mainly on its reactance, the higher the reactance, the smaller the overload capacity of the generator. The reactance of a generator is the product of its armature inductance and frequency. At a certain frequency, the smaller the armature inductance, the smaller the reactance, and vice versa.

Rare earth permanent magnet motor has high power density and low reactance because rare earth permanent magnet has high magnetic energy product and its permeability is close to that of air permeability. If rare earth motor adopts no slot or even no core structure, it can not only reduce the core loss, but also significantly reduce the inductance of the motor and increase the overload multiplier of the motor. Since no slot or even no core motors cannot change the magnetic field of the motor to adjust the output voltage of the motor, such motors should be driven by auxiliary engines.

In order to prevent the short-circuit current from being too high when the high overload PMG is short-circuited, a core inductor can be strung into the armature winding of the motor to limit the current when a short-circuit occurs, and the core inductor is short-circuited by a bidirectional switching device when there is no fault.

In order to reduce the armature copper loss of slotless or coreless motor, the air gap magnetic field of the motor is desirable to be sinusoidal, so the permanent magnet motor composed of Halbach's magnetic ring is used to achieve the strongest magnetic field with the smallest amount of magnet structure. The foreign proposed PMG prototype has a rated power of 5 MW, voltage of 1 000 V, speed of 18 000 r/min, and motor weight of 241 kg.

9.2.3　Superconducting Power Generation Method

The development of high-temperature superconducting materials has created the conditions for the application of superconducting generator. Superconducting generators are characterized by high electromagnetic load, high power density and high overload capacity. However, superconducting generators must have low-temperature medium cooling and an operating temperature lower than the critical temperature of the material.

9.3　Aircraft Actuator Regenerative Energy Absorption

With the increase in the equipment of aircraft with electrical energy, a large number of actuating mechanisms, such as flight control systems, landing gear systems, various pumps and valves, etc. , have appeared in MEA, most of which work with electric motors to drive the actuators. For example, the B787 has high-power speed-controlled motors powered by variable frequency alternators through Auto Transformer Rectifier Unit(ATRU)converter, and there are usually four ways to deal with renewable energy, namely resistance consumption method, energy storage equipment method, energy storage method of airborne special equipment and energy absorption method of driving mechanism.

1. Resistance Consumption Method

Since the ATRU is a one-way electrical energy converter, the regenerative energy of the motor can only be consumed by the resistor in the regenerative energy absorption circuit.

This method is inefficient, and the heat generated by the resistor needs to be dissipated, raising the ambient temperature and burdening the environmental control system.

2. Electricity Storage Equipment Method

The use of power storage equipment to store regenerative energy is a highly efficient method. The energy when the actuator brakes and decelerates is fed back to the storage device, and the energy when it starts to accelerate can be supplied by the storage device.

3. Energy Storage Method of Airborne Special Equipment

Instead of using special power storage equipment, regenerative energy is stored with equipment already on board the aircraft. The electric fuel pump, fan motor, and ring-controlled compressor motor are all existing equipment on the aircraft. When the speed of the electric pump, electric fan, or ring-controlled compressor motor is increased while the renewable energy is being fed back, their rotating parts will absorb the energy, and the increased speed of the pump, fan, and compressor will cause a corresponding increase in the power consumed by them to absorb the regenerative energy. During the non-feedback period, the speed of the pump, fan and compressor gradually decreases, and the mechanical energy stored in the rotor is consumed by the pump, fan and compressor.

4. Actuator Feed-back Energy Absorption Method (S/G Motor Operating Condition)

The S/G works in the motor state with the help of the S/G to absorb the energy returned by the actuator. The S/G can be driven by the main engine, by the APU, or by the Integrated Power Unit(IPU). Since the engine rotation direction of the S/G is unchanged, the torque direction of the S/G is opposite direction when starter motor and generator operation, if the engine drives the motor through the gear box, then due to the clearance between the gears, the forward and reverse motor torque will lead to mechanical shock between the gears, which will easily damage the transmission gears, so this method is suitable to be realized by an internally mounted S/G without gear transmission mechanism.

When the actuator brakes and decelerates to return energy, the AC/DC energy return causes the S/G to enter electric operation to absorb regenerative energy and prevent the bus-bar voltage from rising.

As shown in Fig. 9. 3. 1, it is the system structure diagram to enhance the peak load handling capability of the aircraft PSS, in which S/G 1 and S/G 2 are the S/Gs driven by the main engine, IPU S/G is the S/G driven by the combined power unit, and the load is a load with actuating characteristics, using positive and negative line double line feed. The output of IPU S/G is fed to the load via diodes VD 1 and VD 2. The output of IPU S/G is fed to the load via diodes VD 1 and VD 2, so the two main S/Gs are not running in parallel.

If the diode in Fig. 9. 3. 1 is changed to a bidirectional electronic switch, the IPU S/G can also absorb the regenerative energy of the actuator. Due to the short duration of the peak return energy of the actuator, a bi-directional fast charging electronic switch is necessary, with the advantage that the IPU generator provides or absorbs only transient power.

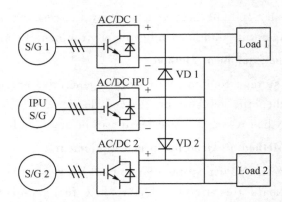

Fig. 9.3.1　System structure diagram to enhance the peak load handling capacity of the aircraft PSS

The third and fourth methods of absorbing regenerative energy are two highly efficient methods and do not require additional storage equipment on the aircraft. However, in the fourth method, the S/G must work with a bi-directional AC/DC converter, which cannot be used to absorb regenerative energy for a brushless DC generator, such as a three-stage synchronous generator and a diode rectifier filter circuit.

9.4　MEA Thermal Management and Energy Optimization

The heat generation and heat dissipation of MEA is different from that of conventional aircraft, mainly reflected in the change of fuselage materials, replacing alloy materials by composite materials, and the change of electrical and electronic equipment equipped with the aircraft. Therefore, it is necessary to conduct thermal management research and energy optimization design.

9.4.1　Impact of Aircraft Body Materials and Equipment

1. Aircraft Fuselage Material

Since the aircraft fuselage is made of composite materials, the thermal conductivity decreases significantly compared with metallic aluminum alloy, and in order to improve the aerodynamic performance of the aircraft, it is required to reduce the aircraft opening and introduce less ram air, so the effectiveness of cooling the in-flight equipment with the help of ram air is constrained. For hypersonic speed aircraft, the thermal effect of airflow increases the temperature of the aircraft fuselage and deteriorates the heat dissipation conditions of the in-flight equipment.

2. On-board Electronic and Electrical Equipment

As the electrical load increases, the avionics equipment has limited efficiency and generates a lot of heat when operating. Similarly, the aircraft power electronics, aircraft generators and motors, aircraft power equipment operating with losses become heat sources.

Transient work of high-powered equipment, instantaneous heat generated more, if not promptly distributed away, will lead to the failure of the equipment itself, surrounding equipment. Therefore, thermal management equipment becomes one of the keys to the reliable work of the equipment in the aircraft.

For decades, storage batteries have been used as emergency power sources, operating for short periods of less than 30 minutes as an energy source. However, with the development of high-voltage DC power supply and the birth of high power and high energy density lithium-ion batteries, the feasibility of parallel power supply of batteries and generators has gained attention. Due to the small internal resistance of the storage battery, and equivalent to a large capacitor, the storage battery has become a good transient energy storage and release device, playing the dual role of emergency power and energy storage.

3. Thermal Storage Materials

The thermal storage unit absorbs a large amount of transient thermal power to prevent over heating of the heat generating equipment and to reduce the peak power of the cooling system. Most heat storage units have a phase change material composition. Paraffin wax is a class of materials with a low phase change temperature that is suitable for use as a thermal storage element. However, the shortcoming of paraffin wax is that the thermal conductivity is very low. In order to improve the thermal conductivity, carbon fibers are placed in paraffin wax to form a hybrid carbon brazing dimension, which increases the thermal conductivity by about 1 000 times and becomes a more desirable thermal storage material.

Taking the B787 as an example, the electric environmental control system is mainly the control of the pressurized cabin, including the environmental control of the passenger cabin, cockpit, electronic equipment cabin and cargo cabin, the cooling of power electronics, feedback electronics and galley equipment, etc. On the one hand, whether it is the regulation of temperature, humidity and cabin pressure is a slow changing process, and the internal temperature/time constant of the aircraft is quite large from the temperature regulation alone, but for transient operating equipment, which generates a lot of heat in a short period of time without allowing the operating temperature to be too high, using the conventional environmental control system's B787 is not a good method. But on the other hand, the environmental control system of the B787 aircraft is suitable for the B787 aircraft, which is far better than the conventional aircraft with the environmental control system with engine compressed air.

The thermal management system of the next generation MEA includes the temperature control system in the environmental control system and the transient thermal energy management of high-power transient overload equipment. It must reasonably mobilize the cooling media on the aircraft, such as fuel, ram air, electric fans and cooling turbines in the air circulator, so that the various equipment of the aircraft can be reliably allowed within the allowed operating temperature range with minimum energy consumption.

9.4.2　Energy Optimized Aircraft

　　Since fuel is the only primary energy source for the aircraft, energy optimization for the aircraft with the goal of consuming the least amount of fuel throughout the completion of the intended mission is the future direction of advanced aircraft development. A schematic diagram of conventional aircraft energy use is shown in Fig. 9.4.1.

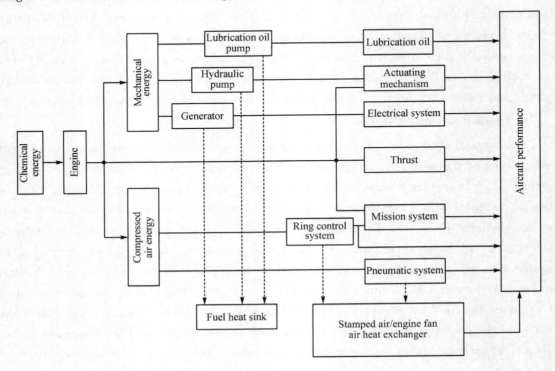

Fig. 9.4.1　Schematic diagram of conventional aircraft energy use

　　Fuel is used to generate thrust through the engine, creating mechanical and pneumatic energy. The mechanical energy makes the slip oil pump, hydraulic pump and generator work. The slip oil is used to lubricate the engine bearings, the hydraulic pump converts the mechanical energy into hydraulic energy to drive the actuating mechanism to operate the aircraft rudder and other machine parts, and the generator converts the mechanical energy into electrical energy to work with electrical equipment.

　　The compressed air extracted from the engine pressurizer provides energy for the aircraft environmental control system and the aircraft air pressure system. Gearboxes, slip pumps, hydraulic pumps and generators all generate losses and heat when they operate, as does equipment that works with electrical energy. Heat is often dissipated using fuel, ram air and air heat exchanges on the inside walls of engine fan intakes to improve the operating conditions and performance of the equipment.

　　Compared with the B787 and B777 aircraft, the most important feature is that the compressed air of the engine compressor is no longer extracted, simplifying the structure of the engine and significantly improving the efficiency of energy use. As shown in Fig. 9. 4. 2, it is a block diagram of the energy use of the MEA, no longer extracting compressed air from the engine.

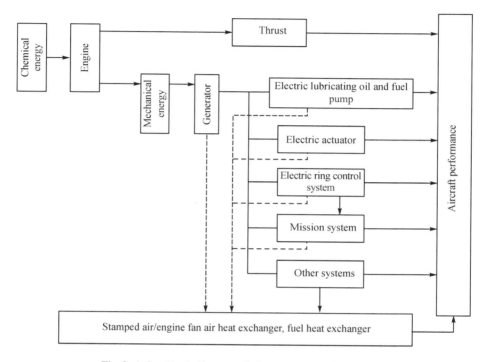

Fig. 9. 4. 2   Block diagram of the energy use of the MEA

No matter how the hydraulic pump, generator, electrical equipment work loss, loss will heat, it needs to cool.

Therefore, improving the efficiency of generator power grid and electrical equipment is another important topic for energy optimization aircraft.

As some equipment on the aircraft is intermittent operating system, short-time operating system, short-time repetitive operating system, etc. , the peak power/rated power ratio is large, operating time/operating cycle ratio is small and other characteristics, the problem of high-efficiency operating point design arises. If the design is designed according to the rated state, the peak operating time cannot meet the requirements, and if the design is designed according to the peak operating time, it will cause great waste, so it must be optimized to seek the best design solution to meet the conditions.

This shows that energy optimized aircraft is an energy and heat optimization system from aircraft, engine, secondary energy to aircraft electrical system, a big system, and the modeling, performance analysis and optimization, energy and thermal management of the system will be on the agenda as one of the strategies for next generation aircraft to be economical, safe and friendly with the environment.

9.4.3　Introduction to Thermal Modeling of MEA Electrical Systems

Due to the complexity and size of aircraft electrical systems, it is of more theoretical importance if we can rely on accurate modeling and simulation tools for analysis and design.

1. B787 Electrical System Structure

The main components of the MEA B787 electrical system, as shown in Fig. 9.4.3, are as follows. Power generation level is:

① Variable frequency AC power with input from engine driven generators, four generators with output of variable frequency 230 V AC, 360—800 Hz and two APU driven variable frequency AC generators.

② 115 V AC, 400 Hz constant frequency AC power, obtained from a 230 V AC busbar via an AC/AC converter, to supply conventional constant frequency AC power loads.

③ 270 V DC, the 270 V DC busbar obtained from the conversion of the main power supply ATRU converter, the main object of the 270 V DC busbar power supply is various motor loads.

④ 28 V DC low voltage DC power supply, 28 V DC generated by 270 V DC via DC/DC converter for battery charger load.

⑤ The conventional inverter VF AC load is connected directly to the main grid.

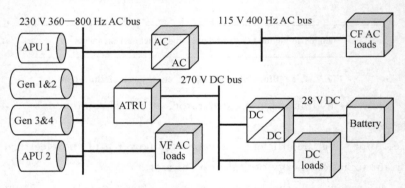

Fig. 9.4.3　B787 electrical system structure

2. Component Model

The main components of the aircraft electrical system are generators, exciters, power converters, batteries, transformers and electrical loads, etc. The electrical component models need to be considered for power loss calculations, the ECS due to heat generation from electrical losses needs to consider factors such as thermal insulation, and the engine shaft must withstand the load moment from the generator.

(1) Power Consumption of Electrical Systems

The electrical load distribution and power conversion efficiency of B787 under cruise, as shown in Fig. 9.4.4, can provide information and basic parameters for modeling.

The total power from the generator in the figure is 944 kW, which is transformed by the

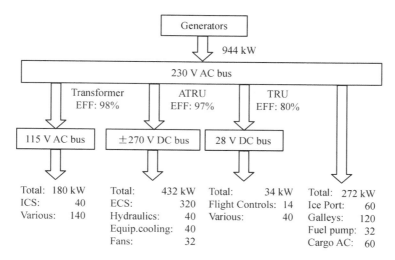

Fig. 9. 4. 4   **Typical electrical system loads and efficiency for B787 (at Cruise)**

transformer to produce 115 V AC busbar, of which the transformer efficiency is about 98%, to supply the conventional load with a total capacity of 180 kV • A; 270 V DC from the inverter AC mains power converted by ATRU, the efficiency of ATRU is about 97%, the total power of the ATRU load is 432 kW, of which 320 kW for environmental control system, 40 kW for hydraulic system, 40 kW for equipment cooling and 32 kW for fan; 28 V DC converted by AC/DC from inverter main AC busbar, with 80% efficiency, total power is 34 kW, 14 kW for flight control, 40 kW for other power-using equipment; power obtained directly from main grid 272 kV • A, 60 kV • A for anti-icing, 120 kV • A for galley, 32 kV • A for fuel pump and 60 kV • A for cargo bay equipment AC.

(2) Information Interaction of Electrical Systems

As shown in Fig. 9. 4. 5, the electrical system is shown in relation to the signals of the engine, APU, ECS, fuel, power distribution system, common load and temperature.

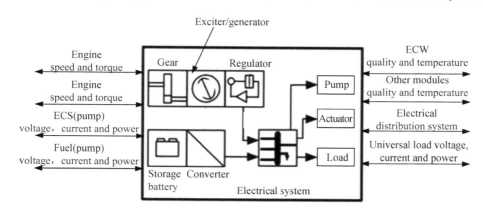

Fig. 9. 4. 5   **Input/output properties of the electrical system**

The signal relationship between the electrical system and the main engine and APU engine, the main signals are the speed and torque signals of the main engine and APU; the signal relationship between the electrical system and the mass and temperature of the environmental control system; the signal relationship between the electrical system and the distribution and general purpose loads, the main parameters are voltage, current, power, torque and temperature.

(3) Generators

The input/output system of the generator is shown in Fig. 9. 4. 6. The input signal comes from the engine speed, which is adjusted by the transmission gear ratio between the engine and the generator, and the terminal voltage and line current of the storage battery are the excitation source signals for the generator. The output of the synchronous generator is a sine wave, and the calculation amount is large.

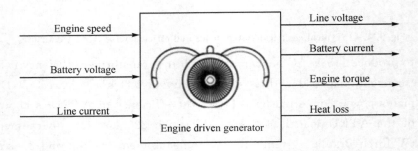

Fig. 9. 4. 6   Input/output system of the generator

The generator line voltage, current output from the battery, engine torque, and losses are the output signals, and the synchronous generator ontology model is shown in Fig. 9. 4. 7. Usually coordinate transformation is used to transform ABC coordinates to dq0 coordinates by PARK to get constant steady-state conditions, larger solution step size and faster calculation speed. Please consult and refer to related literature for details.

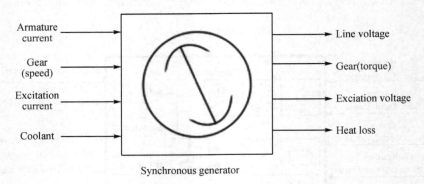

Fig. 9. 4. 7   Synchronous generator ontology model

The electrical component model also takes into account power loss calculations, as well as factors such as insulation of the ECS system.

(4) Exciter

As shown in Fig. 9. 4. 8, it is the block diagram and signal flow diagram of the generator's exciter. Asynchronous generator needs a magnetic field to be excited, and the one that provides this field is called the exciter, usually a wire-wound motor or a permanent magnet synchronous generator, and the shaft of this exciter is co-axial with the main generator shaft.

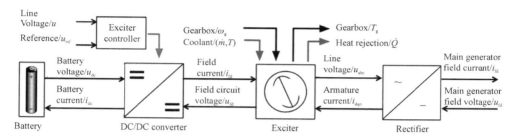

Fig. 9. 4. 8　Block diagram and signal flow diagram of the generator's exciter

The output voltage from the exciter is rectified and delivered to the excitation winding of the main generator. The excitation current must be generated independently of the generator, which is usually provided by a dedicated circuit, and can be controlled and regulated to regulate the voltage at the output of the generator. The excitation current for the exciter can be provided by the storage battery, which is then regulated by a DC/DC converter and supplied to the exciter.

(5) Inverter

An inverter converts a DC voltage to an AC voltage and connects it to the rest of the AC system either directly or through a transformer. To improve simulation speed, the inverter can be modeled using an average switching power loss modeling approach. The main power switching device of the inverter can be equated to an ideal switch in series with a forward voltage drop and a voltage drop across the on-resistance, and its main switching device, the IGBT, will incur conduction losses when it is turned on. A thermal-capacitive model can be used to represent the total temperature of the inverter. It is assumed that the inverter is placed in an electronic compartment with natural convection cooling on its surface.

(6) Rectifier

The rectifier block converts the AC voltage to DC voltage and has an interface with the rest of the DC system. It is an active rectifier circuit, equivalent to a voltage source circuit, and the rectifier circuit usually comes with a low-pass filter.

(7) Storage Battery

The B787 uses a lithium-ion battery, and its mathematical model can be determined from the test values of the specific lithium-ion battery. The equivalent model for the battery can be generic and applicable to many types of batteries, such as lithium-ion, nickel-cadmium, and lead-acid batteries. The circuit model of the battery can be estimated and calibrated using battery test data.

The capacity of the battery is related to the charge/discharge rate, the operating temperature and the number of cycles. The charge/discharge rate is closely related to the magnitude of the current, and the charge/discharge rate can be used to determine the operating condition of the battery. The dynamic capacity expressed by the State Of Charge (SOC) of the battery is a function of these factors as well.

The electrical equivalent circuit diagram of the battery model is shown in Fig. 9. 4. 9. The battery model can be considered using the multi-scale time concept, using seconds, minutes and hours time intervals to build the model, i. e. , the equivalent circuit model for each time scale in the electrical equivalent circuit of the battery can be viewed as a parallel form of each resistor R_{sec}, R_{min}, R_{hr} and capacitor C_{sec}, C_{min}, C_{hr}.

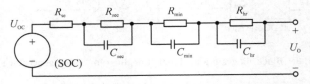

Fig. 9. 4. 9   Electrical equivalent circuit of the battery model

In Fig. 9. 4. 9, U_{oc} is the output open circuit electromotive force of the battery in the charged state, and U_o is the output terminal voltage of the battery.

(8) Transformer

A transformer transforms the voltage of the primary winding into another voltage value of constant frequency on the secondary winding. The winding is insulated from each other and their steady-state electrical model includes the resistive voltage drop (without considering the change in resistance). Mono-phase or multi-phase transformers are analyzed in the same way, and the magnetic coupling of the transformer is ideal, usually considering only the copper losses.

(9) Electrical Load

The electrical load analysis is divided according to the different power levels, current magnitudes or the nature of the impedance, and is seen as a concentrated thermal load for simulation when performing energy calculations and is no longer differentiated. In case of a balanced three-phase system, the steady-state value of the current on the neutral line is considered to be zero.

9.4.4   Power Consumption of Main Flight Profile

The electrical power used by the B787 generators (main flight profile under maximum icing conditions) is shown in Fig. 9. 4. 10, with the total electrical load of all generators for different flight phases. The B787 has installed generators with a total capacity of 1 MV · A, so the maximum icing condition is close to the maximum electrical output of the generators.

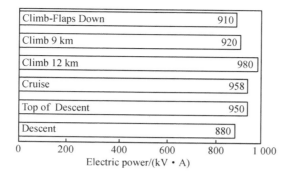

Fig. 9. 4. 10   Electric power of the generator (B787 main flight profile under maximum icing conditions)

1. Electricity Consumption Power Situation

Here, we mainly consider the power used in the climb stage, cruise stage and descent stage, and the power used in other stage such as loading, taxiing and unloading stage is small and is not considered.

(1) Climb Stage

From Fig. 9. 4. 10, it can be seen that the power used in the flap-down state of the climb stage is about 910 kV · A, the power used in the climb to the 9 km position is about 920 kV · A, and the power used in the climb to 12 km altitude is about 980 kV · A.

(2) Cruise Stage

The power used in the cruise stage is close to 950 kV · A, mainly to complete some tasks flight, such as civil airliners to prepare food and other electrical equipment for the kitchen, flight control actuating surface manipulation for micro-adjustment, the power used is less.

(3) Descent Stage

When the descent starts from the highest place in the air, the increased power consumption is mainly used for the control of the flight control surface, with an electrical power of about 950 kV · A. As the descent process proceeds, the power consumption gradually decreases by about 880 kV · A, also for flight attitude adjustment.

2. B787 Electrical Power Status of Each Busbar

As shown in Fig. 9. 4. 11, the power consumption of the B787 at each voltage busbar during 5 flight hours can be roughly divided into 28 V DC, 270 V DC, 230 V AC and 115 V AC for electrical loads. The power consumption of various types of busbar voltages is shown in the curve in Fig. 9. 4. 11. It can be seen that the power consumption of 270 V DC is the most, mainly used for environmental control, flight control, equipment cooling, fan power and other high-power power consumption occasions.

For the purpose of multiple generator power supply equalization loading test, it is assumed that generators 1 and 2 provide power for the loads on the 230 V AC and 115 V AC busbars, while generators 3 and 4 provide power for the loads on the 230 V AC, 270 V DC

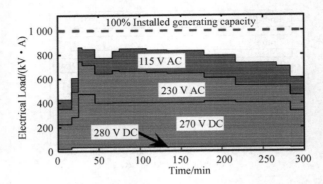

Fig. 9. 4. 11   B787 power usage by voltage busbar（within 5 flight hours）

and 28 V DC busbars, as shown in Fig. 9. 4. 12, for each of the B787 main generator power supply, which gives the power consumption envelope for 5 flight hours curve.

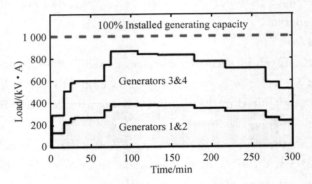

Fig. 9. 4. 12   B787 electrical power supply for each main generator（within 5 flight hours）

It can be seen that the total power used by all four main generators did not exceed their rated value of 1 MV · A, and reached the maximum flight power during the cruise stage between 1 and 2 hours of flight due to the use of some high-powered power-using equipment. In the second half of the flight, the total power generated by the generators gradually decreases as some high-powered power-using equipment is turned off.

3. Generator Temperature Change Situation

As shown in Fig. 9. 4. 13, the temperature variation profiles for generators 1 and 3 were simulated, with slightly different profiles during the 5 flight hours due to the unequal loads between the two generators. Active control of the thermal management system of the generators is not included in these results, thus allowing the temperature of the generators to fluctuate significantly during the simulation. If the thermal management system is included, the temperature of the generators can be maintained within the set thermal constraints.

The integrated electrical system model is used to allow the prediction of the thermal

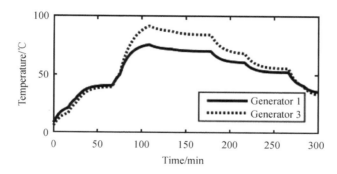

Fig. 9. 4. 13   Temperature variation profiles of generators 1 and 3 during the 5-hour flight

behavior of the integrated components. The main electrical components in the B787 electrical system include the generator, power converter, battery, transformer and load. For the three-phase generator, a coordinate transformation method is used to transform the ABC three-phase coordinates into a dq coordinate system. For the power converter, an average switching modeling method can be used to maintain the necessary voltage, current and power.

The thermal model of each electrical component can be used for temperature monitoring and hot spot detection of generators, power converters and batteries, etc. , for experimental validation in future studies of aircraft electrical systems.

9.5 Summary

The MEA has undergone fundamental changes in the types of electrical loads due to the generation and regulation of electrical energy, and various new technologies have emerged to change the characteristics of energy changes in the electrical system of MEA. Along with the MEA, variable frequency alternator VFAC, SSPC and other high efficiency electrical and electronic equipment in the aircraft electrical system, the constant speed transmission device has been abolished and the power of the main generator has increased significantly while the structure is simple and the power density has increased. The energy-optimized aircraft is the goal of the next-generation MEA. Highly efficient use of aircraft fuel, rational design of aircraft components, subsystems and systems, improving the efficiency of equipment and systems to reduce losses, reducing heat generation, contributing to the work of the equipment itself and the reduction of the volume and weight of the cooling system, and contributing to the further improvement of aircraft performance.

It is estimated that compared with the B777 aircraft, if the new technology of advanced aircraft is adopted, there will be substantial reduction in fuel consumption, pollution emission, noise reduction, aircraft size, weight and other aspects, while the reliability is greatly improved. Electrical technology and aviation technology penetrate each other and promote each other to overcome difficulties for the benefit of mankind.

9.6　Exercises

1. What is regenerative energy?

2. What are the usual ways to handle renewable energy?

3. What are the characteristics of an energy-optimized aircraft?

4. Why does an aircraft consume more power to descend from the air than during the cruise phase?

5. Which power supply usees the most electrical power on the B787? What load is it mainly used for?

Chapter 10 Common Core System of MEA

10.1 Introduction

With the continuous application of computer technology in the aircraft, self-detection, self-diagnosis, networking and other systems have penetrated into the various systems of the aircraft. Traditional aircraft systems usually have dedicated Line Replaceable Unit(LRU) to complete their various tasks. The B787 represented by MEA adopts Common Core System (CCS) to provide common processing and Common Data Network(CDN) to support the operation of many aircraft systems to improve reliability, lower cost and lower weight. As shown in Fig. 10.1.1, it is the CCS in B787.

Fig. 10.1.1 CCS in B787

Fig. 10.1.2 shows the block diagram of the CCS, which only shows the functional connections between the system components. The CCS has many software applications that do calculations for airplane systems.

According to Fig. 10.1.2, the CCS of B787 has the following:

① Two Common Computing Resource (CCR) cabinets, each cabinet has sixteen modules.

② Six ARINC664 Network Remote Switches(ARS).

③ Twenty-one Remote Data Concentrates(RDC).

All modules use Fiber Optic Bus(FOB) and Electrical Bus (EB) for signal interaction.

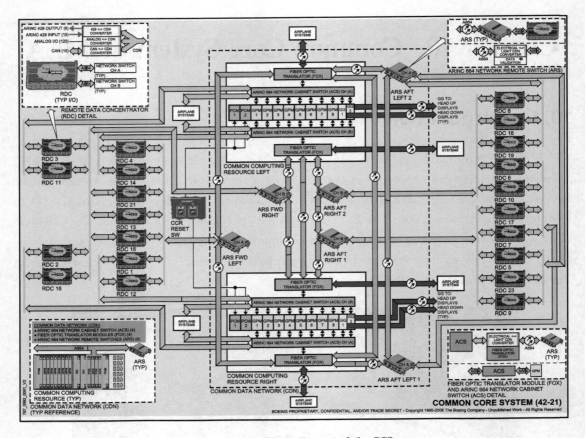

Fig. 10. 1. 2　Block diagram of the CCS

10.2　Common Computing Resource

As shown in Fig. 10. 2. 1, the CCR cabinets are located in the forward Electronic Equipment(EE) bay. The two CCR cabinets contain the main components of the common core system.

Fig. 10. 2. 1　CCR cabinets（in forward EE bay）

The CCR cabinets are located in the forward electronics bay: the left CCR cabinet is at the bottom of the E - 1 rack, the right CCR cabinet is on top of the front wheel well box. The composition of these cabinets is exactly the same, and the placement position is set by the program. Each CCR cabinet can be connected to the busbars of the aircraft's primary and secondary power, or to the front electronics cooling system, or with 10 Mb/s electrical and 100 Mb/s fiber ports. Each CCR cabinet has 16 slots for line replaceable units. As shown in Fig. 10. 2. 2,it is the CCR line replaceable module slot.

Fig. 10. 2. 2　CCR line replaceable modules slots (16 slots)

10.2.1　CCR Cabinet-Line Replaceable Modules

Fig. 10. 2. 3 shows the line replaceable modules in the CCR cabinet, with two Power Conditioning Modules (PCMs) on both sides of the CCR cabinet, 8 General Processor Modules (GPMs), 2 Graphics Generators (GGs) modules, 2 ARINC664 network Cabinet Switches (ACS), 2 Fiber Optic Translators (FOXs) modules. All modules of the same type are interchangeable, the General Processor Module (GPM) must have software installed correctly.

Fig. 10. 2. 3　Line replaceable modules (in CCR Cabinet)

1. CCR Cabinet Fan/Valve Assemble, and Cooling Duct

As shown in Fig. 10. 2. 4, it is the cooling fan and valve, the fan and valve components are located on the back of the CCR cabinet.

① The normal cooling air for each CCR cabinet comes from the cooling ducts of the E/E system of the front electronic equipment of the aircraft.

② If normal cooling does not work, the fan and valve assembly on the CCR cabinet will work, the fan will blow ambient air through the CCR module, and the valve will close the E/E cooling duct to prevent the fan from only sucking air from the duct. If the fan and valve assembly fails, the cooling status information for the left and right cabinets of the CCR is displayed.

③ The fan and valve assembly is an LRU, and the fan is powered from the CCR cabinet PCM.

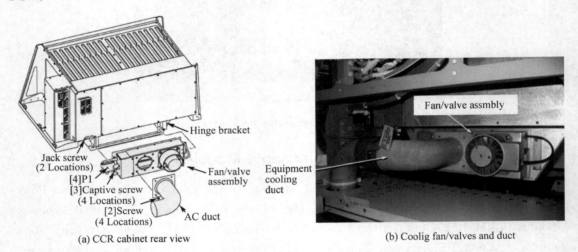

(a) CCR cabinet rear view   (b) Coolig fan/valves and duct

Fig. 10. 2. 4   Cooling fan and valve (in CCR cabinet)

2. Power Conditioning Module (PCM)

① The PCMs make high quality power for the CCR cabinet modules.

② Each CCR cabinet contains 2 PCMs, which are 28 V DC busbars and HBB.

③ HBB provides emergency backup power and supplies power for internal clock and GPM volatile memory.

The PCM only needs a 28 V DC power supply to power the CCR cabinet modules.

3. General Processor Module (GPM)

① GPM has general software processing functions for CCS dedicated software applications.

② Each GPM is a computer with a processor, memory and power supply.

③ The application software performs fault alarm, processing and calculation for the independent aircraft system, including Display and Crew Alert Function (DCAF), Hydraulic

Interface Function (HYDIF), Landing Gear Drive System/Nose Wheel Steering System (LGDS/NWSS), Central Maintenance Calculation Function (CMCF), Flight Management Function (FMF), Data Communication Management Function (DCMF).

④ Use software to perform fault alarm, processing and calculation for independent aircraft systems, including display and crew alert function, hydraulic interface function, landing gear drive system/nose wheel steering system, central maintenance computing function, flight management function, data communication management function.

⑤ Each CCR cabinet contains 8 general-purpose processor modules. As shown in Fig. 10. 2. 5, it is the General Processing Module (GPM) module. The GPM module numbers in Fig. 10. 2. 5(a) are counted from right to left as 1 to 8, Fig. 10. 2. 5(b) shows the GPM module in the common computer resource, and the entire CCR resource exchanges information with other devices through the ARINC664 bus remote switch ARS.

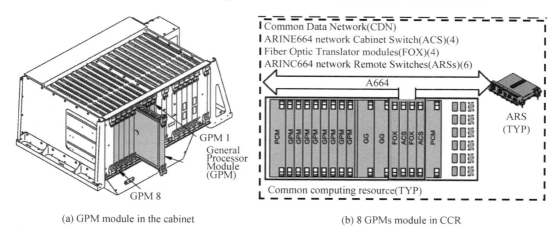

(a) GPM module in the cabinet　　　　(b) 8 GPMs module in CCR

Fig. 10. 2. 5　General processing module GPM module

Each GPM contains two 800 MHz processors, and consists of two single Application Specific Integrated Circuits(ASICs) and two end systems, for a total of four such ASICs. Memory for core operating system software and functional application software is 128 MB Flash (volatile), 512 MB SDRAM (non-volatile) and 32 MB mailbox memory (reserved area for packets) and power monitors.

The status message of the CCR includes the information of the GPM status on the left and right sides, and GPM 1—GPM 8 on each side respectively display the GPM. When the normal operation cannot be completely performed, check whether there is a hardware failure. Only modules in GPM slot 8 can be used for scheduling. If the GPM module needs to be replaced or moved, the software must be reloaded.

10.2.2　ARINC664 Network Switch

CCS with 10 ARINC664 Network Switches for monitoring correctness of operation, sending data to applications such as aircraft systems and GPM modules, converting data formats between optical and electrical signal formats. A network switch has either Channel A or Channel B as the switch operation channel, providing two independent data paths. The network switch is software loadable. There are two types of network switches, namely ARINC664 network Cabinet Switch (ACS) and ARINC664 network Remote Switch (ARS).

1. Network Cabinet Switch ACS

As shown in Fig. 10. 2. 6, it is the network cabinet switch ACS module. The network cabinet switch ACS is used to send/receive the ARINC 664 electrical signal between the GPM module; transmits/receives ARINC 664 electrical signals with the FOX data module, and has interfaces to some aircraft systems with ARINC 664 electrical signals. All ARINC 664 data to and from the CCR cabinet must go through the network cabinet switch ACS. There are 2 network switch channels in each CCR cabinet, channel A and B. The removal/installation of the ACS is the same as for other modules in the CCR cabinet.

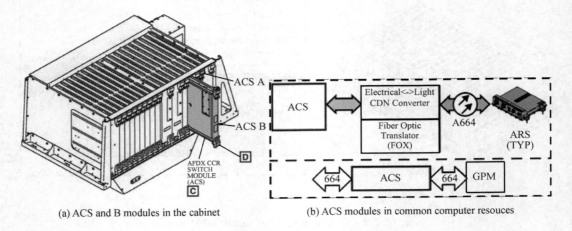

(a) ACS and B modules in the cabinet   (b) ACS modules in common computer resouces

Fig. 10. 2. 6　Network cabinet switch ACS module

2. Fiber Optic Translator (FOX) Module

The two(FOX) modules have the function of translating ARINC 664 electrical data and ARINC 664 optical fiber data to each other, and perform data transmission between the left and right CCR cabinets through the fiber optic bus. Each CCR cabinet has 2 independent data path A and B channel FOX modules. And the two FOX modules in the CCR cabinet are adjacent to the ACS modules. But the FOX module does not contain operating software, so its removal or installation is similar to the ACS and GPM modules. When any FOX module fails to perform data transmission or optical/electrical translation, etc. , the fault information can be displayed in the status information of the CCR, and all 4 FOX modules must be running to process data quickly.

3. ARINC664 Network Remote Switch Module

There are six ARINC664 network Remote Switches (ARS) in CCS. As listed in Tab. 10. 2. 1, they are ARS modules in the CCS.

Tab. 10. 2. 1 ARS modules in the CCS

SN	Left	Number	Channel	SN	Right	Number	Channel
1	Front electronics bay	1	A	3	Front electronics bay	1	B
2	Rear electronics bay	2	A	4	Rear electronics bay	2	B

The six ARINC664 network Remote Switch(ARS) modules of the B787 aircraft are located in different locations of the aircraft, as shown in Fig. 10. 2. 7.

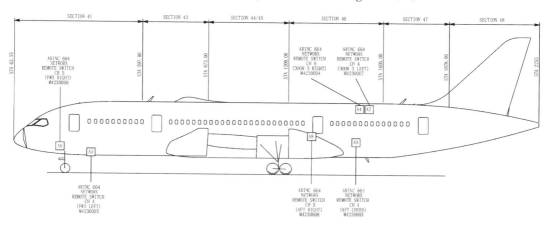

Fig. 10. 2. 7 Location of ARS in the aircraft

As shown in Fig. 10. 2. 8, it is the ARS module. Fig. 10. 2. 8 (a) shows a schematic diagram of data communication with the remote data concentrator module through the ARINC664, and the Fig. 10. 2. 8(b) shows the ARS assemble.

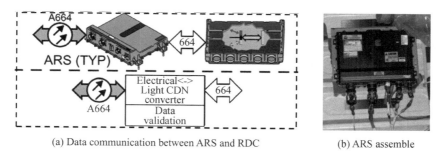

(a) Data communication between ARS and RDC   (b) ARS assemble

Fig. 10. 2. 8 ARS module

The ARS in Fig. 10. 2. 8 is used to receive or send data from the FOX module, transform the FOX signal into electrical signal data, receive or transmit aircraft system data from the remote data concentrator RDC, and change the RDC electrical signal to optical fiber data signal.

10.2.3 Remote Data Concentrate

B787 has a total of 21 Remote Data Concentrates(RDCs), 11 at the front of the aircraft and 10 at the rear of the aircraft. The RDCs are close to the aircraft system components to which they are connected. The layout on the aircraft is shown in Fig. 10.2.9.

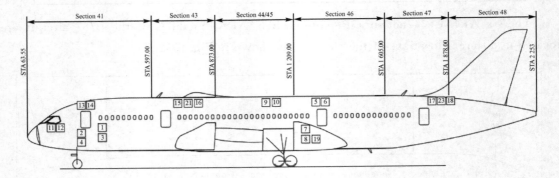

Fig. 10.2.9 Location of RDCs on the aircraft

Tab. 10.2.2 lists the positions on the aircraft corresponding to the 21 RDCs. In order to more easily connect with the various system components of the aircraft, the RDCs are scattered in the front and rear of the aircraft. There are mainly, FWD E/E bay and AFT E/E bay, No.1, No.2, No.3 and No.4 doors and the cockpit.

Tab. 10.2.2 Distribution of 21 RDCs

SN	Number	Position	SN	Number	Position	SN	Number	Position
1	RDC – 1	FWD EE bay	8	RDC – 8	AFT EE bay	15	RDC – 15	Door 2
2	RDC – 2	FWD EE bay	9	RDC – 9	AFT EE bay	16	RDC – 16	Door 2
3	RDC – 3	FWD EE bay	10	RDC – 10	AFT EE bay	17	RDC – 17	Door 4
4	RDC – 4	FWD EE bay	11	RDC – 11	Cockpit	18	RDC – 18	Door 4
5	RDC – 5	Door 3	12	RDC – 12	Cockpit	19	RDC – 19	AFT E/E bay
6	RDC – 6	Door 3	13	RDC – 13	Door 1	20	RDC – 21	Door 1
7	RDC – 7	AFT EE bay	14	RDC – 14	Door 1	21	RDC – 23	Door 4

RDC is the I/O interface between network switches and aircraft systems. RDC receives and sends aircraft system data, namely analog signals and analog discrete signals, ARINC429 bus data, and CAN bus.

RDC converts the aircraft system data into ARINC664 format and sends it to the network switch; ARINC664 data received from the network switch can be converted into analog signal, ARINC429 signal or CAN bus data and sent to the aircraft system.

Some RDCs have interfaces with network cabinet switches ACS. RDCs only send and receive electrical signals (no fiber connections). Each RDC has 2 channels (A and B). The A channel is connected to the A channel network switch, and the B channel is connected to the B channel network switch.

As shown in Fig. 10. 2. 10, it is the RDC, in which Fig. 10. 2. 10(a) shows the data communication between the RDC and the network switch through channels A and B respectively, and Fig. 10. 2. 10(b) shows the remote data concentrators assembly.

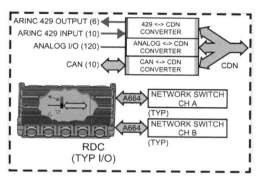

(a) Connection of RDC to network switch   (b) Remote data concentrators assembly

Fig. 10. 2. 10   Remote data concentrator

The CAN bus is a digital serial bus that allows multiple data points on the bus. For example, each fire detection device in a fire detection system can transmit fire alarm data to the CAN bus. The fault of the RDC can be released through the status information of the CCS. There are 21 RDCs in the B787. If the RDC fails, it can be sent to the CCS through the ARS.

10.3   Common Data Network Applications

As shown in Fig. 10. 3. 1, the CDN distance measuring equipment system consists of the followings:

① Two Distance Measurement Equipment ANTENNAs(DME ANTENNAs).

② Two distance measuring transponders.

③ An Audio Gateway Unit (AGU).

④ Two Remote Data Concentrates (RDCs).

⑤ ADME ground station.

⑥ An Audio Control Panel (ACP).

⑦ A Common Computing Resource Cabinet(CCRC).

⑧ Ground maintenance computer.

⑨ MFD,keypad and cursor control device.

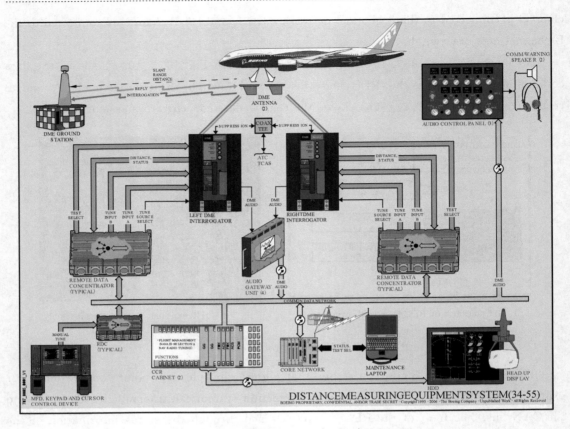

Fig. 10. 3. 1   CDN distance measuring equipment system

There are ACS, ARS, FOC modules and the connection between them in the CDN, and the CDN in the CCS has six ARINC 664 network switches ARS modules. Fiber optic connection is used between ARS and CCR cabinets, and there are 2 ACS and 2 FOX modules in each CCR cabinet.

1. Power Supply Common Core System

As shown in Fig. 10. 3. 2, it is the CCS power start operation panel. When the power is turned on, the CCS will automatically start the power supply. There are 2 types of CCS power-up modes, namely uninhibited mode and inhibited mode. In the uninhibited mode, the GPMs do a full Power-up Built In Test (PBIT). The test takes 3 minutes. Hosted functions are not available until the test is complete. In the inhibited power-up mode, the GPMs do not do a full PBIT. Hosted functions are available in about 50 seconds.

The CCS starts in the inhibited mode when any of these conditions exist: The airplane is in the air, a fuel cutoff switch is in the RUN position, the airplane is on battery power only, any of the above data is not available. Network switches, FOX modules, and RDCs always start in less than 50 seconds. The power-up mode is complete when the displays show their default formats.

As shown in Fig. 10. 3. 3, it is the CCR cabinet reset switch. On the operation panel, there are battery switch, tow power switch, hatch switch, CCS switch (one on each side), emergency lighting switch, oxygen switch, window heating backup switch, etc.

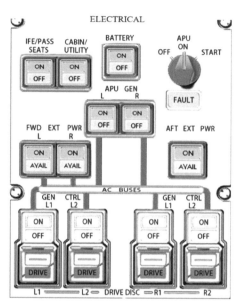

Fig. 10.3.2　CCS power up

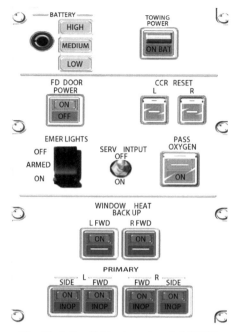

Fig. 10.3.3　Cabinet reset switch(CCS)

Using these switches allows the Power Conditioning Module(PCM) on the side of the CCR cabinet to be recycled and can be connected directly to the PCM. These switches are used by the crew when the display is not showing.

2. Common Core System's Maintenance Page Guide

As shown in Fig. 10.3.4, it is the CCS Maintenance page, which displays data about the status of CCS components.

The main contents to be displayed on the CCS maintenance page include system page, maintenance selection, maintenance page menu display, etc. You can also select the bottom of page 2 on the right scroll bar, and you can see CCS that number 42 for selection.

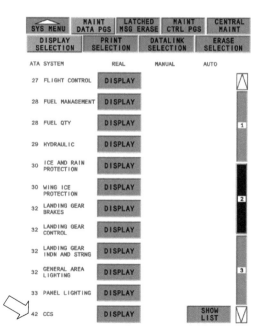

Fig. 10.3.4　CCS Maintenance page

10.4　Summary

Compared with conventional aircraft, MEA has a special Common Core System for data information exchange of each system.

Traditional airplane systems typically have dedicated Line Replaceable Units(LRUs)to do their various tasks. The Common Core System (CCS) supplies a common processing and common data network to support the operation of many airplane systems which provides: improved reliability, lower cost and lower weight.

The CCS has many software applications that do calculations for airplane systems.

The CCS has these components: Two Common Computing Resource (CCR) cabinets, and each cabinet has 16 modules. Six ARINC664 network Remote Switches (ARS). Twenty-one Remote Data Concentrators (RDCs).

10.5　Exercises

1. What are the functions of the common core systems of MEA?
2. What are the functional modules in the common computing resource cabinet?
3. What is the role of the network switches?
4. What is the function of the aircraft remote data concentrator?
5. How many remote data concentrators are there on the B787 aircraft?

Chapter 11　Application of Power Electronics Technology of MEA

The harsh operating environment of aviation electrical equipment has extremely demanding requirements in terms of performance, reliability, weight, volume and energy consumption. Traditional electrical equipment mainly uses power electronics devices made of Si semiconductor, but because of 60 years of development of Si power electronics devices, its performance is close to its theoretical limit, and it is difficult to have a substantial increase, which has become one of the bottlenecks limiting the aviation secondary power supply to further improve performance.

SiC power devices are resistant to high temperatures, radiation, high breakdown voltages and operating frequencies, making them suitable for operation in harsh conditions. Compared to conventional Si power devices, SiC power devices offer a significant reduction in power consumption. As a result, the size and weight of power electronics devices can be significantly reduced, reliability can be improved and their application prospects in MEA are extremely broad.

The US has long set its sights on SiC devices, which are used predominantly in the power conversion and power distribution units of its third generation MEA.

11.1　Challenges for Variable Frequency AC Power System

In the MEA, the use of electricity is increasing, and the airborne Constant Frequency Power Supply System (CFPSS) cannot provide enough power for the increased load of the MEA, so the MEA power system is changing from the constant frequency AC system to the frequency conversion AC system.

MEA not only reduces the complexity of aircraft, but also increases reliability, fault tolerance, power density and performance. Recent developments in the fields of power electronics, electronic drives, control electronics and microprocessors have provided the impetus for improving the performance of aircraft electrical systems and their reliability. Therefore, the concept of MEA is seen as the way forward for aircraft power system technology. As shown in Fig. 11.1.1, the structure of a Constant Frequency Power Supply System(CFPSS) and a Variable Frequency Power Supply System(VFPSS) is illustrated.

As can be seen from Fig. 11.1.1 (a), the constant frequency alternator generates a constant frequency AC power supply for the AC load, and the transformer rectifier converts 115 V AC power into 28 V DC power supply. This PSS is simple in structure, but the

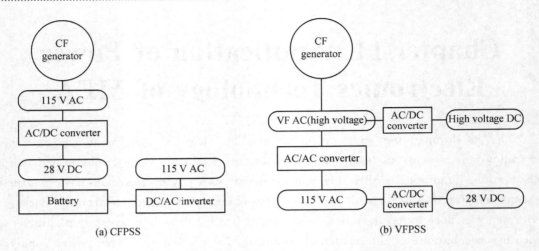

Fig. 11. 1. 1   Aircraft power system structure (one channel)

constant frequency AC power supply of 115 V AC, 400 Hz is not the best input power supply for some AC loads.

Fig. 11. 1. 1 (b) is the frequency conversion power system, the primary power is generated by the frequency conversion generator. The frequency of the AC power supply will continuously vary with the engine speed. For optimal performance and efficiency, the frequency conversion alternator produces high voltage AC power supplies, using a variety of different power electronic converters, such as AC/AC and AC/DC converters, to 115 V low voltage AC and DC power supplies to meet the needs of various loads.

In order to achieve lower weight, smaller size, lower cost and better performance, the VFPSS in MEA requires a special design of the power electronics converter and microprocessor to meet the requirements of light weight, small size and compactness, direct connection to the variable frequency input power, low harmonics and meeting EMC requirements. This poses significant challenges for the use of power electronics in MEAs, the main ones being the following.

11. 1. 1   The Harmonic Challenge

The increased harmonic requirements of various electrical and electronic equipment is one of the greatest challenges. The large number of converters and frequency converters in PSS has led to an increase in harmonic currents in aircraft power grids.

In order to limit the adverse effects of harmonic currents on airborne power generation and distribution systems, the international standards ISO – 1540 and DO – 160 limit the input harmonic current distortion of airborne user equipment, ISO – 1540 was the first requirement to add harmonic requirements, DO – 160 has added harmonic requirements since version D and has almost identical harmonic requirements, Tab. 11. 1. 1 lists the defined limits for input harmonic currents.

Tab. 11. 1. 1   Harmonic requirements in ISO – 1540 and DO – 160F

SN	Number of harmonics	Harmonic limit values	
		Mono-phase	Three phase
1	3	$I_h=0.15I_1/h$	$I_h=0.02I_1/h$
2	5, 7		$I_h=0.02I_1/h$
3	11		$I_h=0.1I_1/h$
4	13	$I_h=0.3I_1/h$	$I_h=0.08I_1/h$
5	17, 19		$I_h=0.04I_1/h$
6	23, 25		$I_h=0.03I_1/h$
7	29,31,35,37		$I_h=0.3I_1/h$
8	Odd times and multiples of 3, 9,15,21,⋯,39	$I_h=0.15I_1/h$	$I_h=0.1I_1/h$
9	2,4	$I_h=0.01I_1/h$	$I_h=0.01I_1/h$
10	Even harmonics greater than 4 6,8,10,⋯	$I_h=0.0025I_1/h$	$I_h=0.0025I_1/h$

Where: I_h is the respective harmonic current and I_1 is the amplitude of the fundamental current (steady state condition)

As can be seen from Tab. 11. 1. 1, the harmonic requirements for the aircraft power supply are very strict. Studies have shown that in order to meet the minimum input harmonic current requirements, a front-end controlled active PWM converter to AC busbar connection scheme is used.

On the one hand, effective control of the input current harmonics to meet the stringent specification requirements remains a major challenge. It is difficult to meet the whole requirement at the same time without reducing the voltage transfer ratio of the converter, the input power factor and the robustness to unbalanced and distorted voltage inputs.

On the other hand, these requirements also present a major design constraint for converter filter design. The use of a relatively small second-order input filter to smooth out the switching ripple current on the power input line in the converter, typically replacing the topology's active front-end control and input power filter design to further reduce weight and size, is also a significant challenge.

11.1.2   Power Factor Challenge

As listed in Tab. 11. 1. 2, they are the requirements for steady-state power factor at full load for motor loads in standard DO – 160F (environmental conditions and test procedures for airborne equipment).

The fundamental frequency of the constant frequency mains supply is 400 Hz and the fundamental frequency of the inverter mains supply is 360—800 Hz. The high fundamental frequency is a great challenge for the power factor to meet the on-board power quality requirements. A wider control bandwidth is required to achieve low input current distortion and high input power factor at high line frequencies. The main source of input current harmonics at high frequency is over-zero crossover distortion, i. e. distortion near the line voltage over-zero point.

Tab. 11.1.2　Power factor requirements for motor loads in Standard DO – 160F

SN	f/Hz	DO – 160 limit value	
		Power factor (lead)	Power factor (lag)
1	>500		>0.7
2	$360<f<500$	>0.98	$2.86\times10^{-3}\times f-0.73$
3	$\leqslant360$		>0.3

The frequency of the VFPSS is mostly higher than the frequency of the constant frequency power system, therefore the voltage drop of the cable in the VFPSS is higher than the voltage drop of the CFPSS, the cable impedance will vary with the AC frequency of the VFPSS, the VFPSS will face more challenges to the PFC, the power factor correction PFC should be designed so that the reactive power varies with frequency.

11.1.3　Other Challenges

The power electronics in the VFPSS must meet on-board EMC requirements, which also present a significant challenge. EMC requirements typically include EMI, High Intensity Radiated Field (HIRF) and lightning. As the converters in VF EPS are high-power device, their performance varies due to the frequency, and they may be difficult to meet the EMC requirements over the entire frequency range.

The size and weight of the unit is a big issue on aircraft due to constraints such as size and weight and cost. Filters can meet harmonic, power factor and EMC requirements, but their size and weight requirements are limited. In addition, AC loads with VF input power need to be selected at the right design frequency to obtain optimum performance, as well as minimum size and weight. Therefore, light weight, compact electronics are another challenge for aircraft applications. There are also environmental condition challenges, such as temperature, vibration, humidity, etc.

11.2　SiC Material Properties and SiC Devices

11.2.1　Overview

As early as 1824, the Swedish scientist J. J. Berzelius discovered the existence of SiC, and although subsequent research further revealed the material's superior properties, the outstanding achievements and rapid development of Si technology at the time diverted the interest of research workers. It was not until the 1990s that the performance bottleneck in Si-based power electronics converters reignited the interest of power electronics researchers in SiC.

Throughout the history of semiconductor materials, Si and Ge are usually referred to as the first generation of semiconductor materials, while GaAs, AlAs and other alloys developed in the 1960s are the second generation of semiconductor materials. High voltage ($>$ 600 V) Si MOSFET on-state resistance by the Si material insulation breakdown field constraints and reached the limit, has been unable to adapt to power electronics in high temperature, high frequency, high power, strong radiation and other extreme environments, and thus following the first and second generation of semiconductor materials, people look to the SiC, GaN and other third-generation WBG high temperature semiconductor materials.

As SiC is developing more rapidly and is more suitable for power conversion, this chapter focuses on SiC materials and the SiC devices made from them.

11.2.2   SiC Material Properties

SiC has a higher breakdown electric field strength, carrier saturation drift rate, thermal conductivity and thermal stability than Si due to the stronger covalent bonds between carbon and silicon than between carbon atoms. SiC has many different crystal structures and more than 250 have been identified. Although there are many lattice types of SiC materials, the only two commercially available are 4H-SiC and 6H-SiC. 4H-SiC has a higher carrier mobility than 6H-SiC, making it the material of choice for use in SiC power electronics devices. The physical properties of several semiconductor materials at room temperature are listed in Tab. 11.2.1.

Tab. 11.2.1   Properties of several semiconductor materials at room temperature

Properties	SiC			Si	GaAs	GaN
	4H – SiC	6H – SiC	3C – SiC			
Band gap width/eV	3.26	3.03	2.20	1.12	1.43	3.39
Insulation breakdown electric field/$(MV \cdot cm^{-1})$	2.20	2.50	2.00	0.30	0.40	2.00
Thermal conductivity/$[W \cdot (cm \cdot K)^{-1}]$	4.90	4.90	4.90	1.50	0.46	1.30
Saturation rate/$(10^7 \times cm \cdot s^{-1})$	2.00	2.00	2.50	1.00	1.00	2.20
Dielectric constant	10.10	9.66	9.70	11.90	13.1	9.00
Electron mobility/$(cm \cdot V^{-1} \cdot s^{-1})$	1 000	500	1 000	1 500	8 500	1 250
Cavity mobility/$(cm \cdot V^{-1} \cdot s^{-1})$	115	40	40	600	400	—

Fig. 11.2.1 shows a comparison of the main physical properties of commonly used semiconductor materials.

From Tab. 11.2.1 and Fig. 11.2.1, it can be seen that the physical properties of SiC and

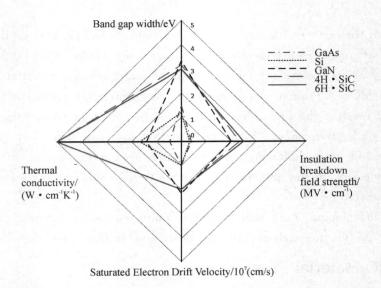

Fig. 11.2.1　Comparison of the main physical properties of commonly used semiconductor materials

GaN semiconductor materials have the following main advantages.

① With a bandgap width three times that of Si, the leakage current of SiC and GaN is greatly reduced and SiC and GaN are both resistant to radiation, making them particularly suitable for aviation and aerospace applications. Theoretically, the operating temperature of SiC power devices can reach 600 ℃, which is much wider than the operating temperature range of Si devices.

② The insulating breakdown field is 10 times that of Si, which greatly increases the voltage capacity, operating frequency and current density of SiC and GaN power devices, while also significantly reducing the on-state losses of the devices.

③ The electron saturation drift rate is 2 times that of Si, which allows SiC and GaN power devices to operate at higher frequency.

④ SiC has 3 times higher thermal conductivity than Si and this gives it excellent heat dissipation and helps to increase the power density and integration of SiC power devices.

It is worth noting that the thermal conductivity of GaN materials does not offer any advantage over that of Si materials.

It can be seen that SiC materials have incomparable performance advantages of Si materials, and its SiC power devices are more suitable for high temperature and high voltage, high power and other harsh environments, and can meet the needs of the rapid development of power electronics technology.

11.2.3　Development of New Wide Band Gap(WBG) Power Semiconductor Devices

From power diodes, which have been used since the early 1950s, to Si MOSFETs,

IGBTs and BJTs, which are now widely available, power semiconductor devices have been widely used in the field of power electronics, and from the point of view of voltage levels and on-resistance parameters of power devices, Si MOSFETs up to 1 200 V and Si IGBTs with higher voltage levels are now available on the market. Under high voltage (>600 V) conditions, conventional Si power devices generally have a large on-resistance problem, which leads to increased losses during use and increased requirements for thermal conditions. The current typical SiC, GaN power device compared to Si power device in the on-state resistance, operating frequency and other aspects of outstanding performance, and with the increase in voltage level, SiC power semiconductor devices in the on-state resistance of this performance has Si power device incomparable huge advantages. From the current development situation, SiC and GaN power devices have greater development potential in high-voltage and low-voltage fields respectively, with voltage levels ranging from 600 V to 900 V.

Due to its outstanding performance advantages, the new WBG semiconductor materials and their power devices are receiving more and more attention. For example, a large number of research institutions and companies (such as CREE, Infineon, ROHM, etc.) at home and abroad have been devoted to the development and production of SiC power device. As can be seen from Fig. 11. 2. 2, the current SiC power devices are still commercially available in the 600 V to 1 200 Voltage range. Of course, there is also research on ultra-high voltage SiC power devices, such as North Carolina State University, CREE Corporation, Powerex, GE, and the Swiss Institute of Technology in Zurich are currently researching high voltage and high power SiC power device around 10 kV.

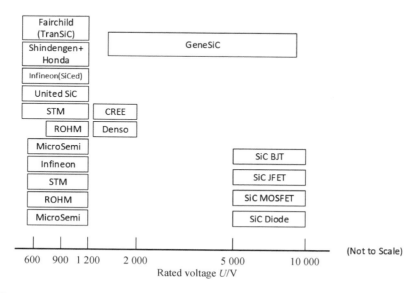

Fig. 11. 2. 2 **Major existing SiC power semiconductor device manufacturing companies**

In addition, according to the SiC power electronics market forecast published by the

French market research firm Yole Development in 2013, SiC power devices will be used across the low-voltage, medium-voltage and high-voltage ranges (300—6 600 kV) and in a wide range of applications, including photovoltaics, wind power, energy-efficient buildings, smart grids, high-voltage power transmission and distribution and power conversion, motor control, hybrid vehicles. The SiC devices are used in a wide range of applications, including photovoltaics, wind power generation, energy-efficient buildings, smart grids, high-voltage power transmission and distribution and power conversion, motor control, hybrid vehicles, rail transportation and aerospace. In these applications, the replacement of Si devices by SiC devices can bring significant performance improvements and economic benefits.

In addition to the outstanding performance advantages of SiC power devices, it is also necessary to take into account their production costs, which are undoubtedly increased by the need to use SiC substrates in the production of SiC power devices. However, with the increasing attention paid to SiC devices at home and abroad, the number of suppliers of substrates has increased and the quality has been greatly improved, which has reduced the production cost of SiC devices to a certain extent, but the production cost of SiC devices is still much higher than that of conventional Si devices. In terms of research to reduce the production cost of SiC power devices, ROHM has publicly stated that the price of existing planar SiC MOSFETs was seven to ten times that of Si IGBTs in FY2013 and is expected to drop to 3 to 4 times in FY2016, while trench SiC MOSFETs are expected to drop to 1.5 to 2 times in FY2016, and CREE has also released low-cost SiC MOSFETs. At the SiC International Society ICSCRM 2013 held in October 2013, CREE has produced power components made from 10 mm^2 6-inch wafers with a qualification rate of 98.7%. In addition, Japanese company Sicoxs also proposed a new method to reduce the production of SiC single crystal substrate, which can reduce the cost of SiC device manufacturing to less than half.

It can be seen that the reduction of SiC power device production cost has become the consensus of many research institutions and device manufacturers at home and abroad, the effective reduction of SiC power devices production cost will be conducive to the popularity and application of SiC power devices.

In addition to SiC power devices, GaN devices, which have just been developed in recent years, offer the same outstanding performance as SiC over conventional power semiconductor devices. Moreover, since GaN power devices can use Si as a substrate, the cost can be reduced to a greater extent compared to SiC power devices, thus achieving a higher cost-performance advantage. Companies such as Panasonic and Sharp have produced GaN power devices with voltage levels as high as 600 V, and the US company Transphorm has produced GaN power devices with a voltage resistance of 600 V on 6-inch Si substrates, and has started to develop and produce GaN power devices with an aperture of 8 inches. However, due to the late development, GaN power device is difficult to reach the voltage level of SiC power devices in the short term, but its development and performance is already enough to cause some impact on low-voltage Si power devices.

11.2.4　Characteristics of Wide Band Gap SiC Power Devices

Below is a brief description of the main characteristics of the four types of SiC power semiconductor devices in relatively mature development.

1. Classification and characteristics of the main SiC power devices

(1) SiC Diode

Low reverse recovery loss, low variation of reverse recovery loss with temperature rise, lower EMI.

(2) SiC JFET

1) Normally-on type

① Voltage controlled normally-on devices, where the power devices are in the on-state under normal conditions and are not easy to use.

② The need to apply a negative offset voltage to ensure reliable switching off of the device, which increases the complexity of the system power supply.

③ SiC JFETs offer higher high temperature reliability than SiC MOSFETs.

2) Normally-off type

① Voltage-controlled normally-off type, where the power devices is in the off-state under normal conditions, for ease of use.

② A higher gate drive current IGS is required to switch on SiC JFETs compared to SiC MOSFETs.

③ SiC JFETs offer higher high temperature reliability than SiC MOSFETs.

3) SiC MOSFETs

① Voltage-controlled normally-off type, where the power devices is in the off-state under normal conditions, for ease of use.

② There are parasitic body diodes.

③ SiC MOSFETs are turned on under conditions similar to normal Si MOSFETs, but require a gate drive voltage of up to approximately 20 V to achieve their optimum performance.

④ The current high temperature reliability of the gate oxide layer is low.

4) SiC BJT

① Normally-off devices with current-controlled, which are not compatible with the voltage-controlled driver circuits currently in use in large numbers.

② There is no secondary breakdown, which is a problem with conventional Si BJTs.

③ The positive temperature coefficient of the on-resistance Ron and the negative temperature coefficient of the current gain make SiC BJTs easy to use in parallel.

2. Comparison of the Characteristics of SiC Power Device in the 1.2 kV Voltage Class

In order to investigate the specific characteristics of SiC JFETs and SiC MOSFETs in depth, the on-resistance characteristics, threshold voltage characteristics, transfer characteristics, avalanche breakdown resistance, switching speed and switching loss characteristics of the devices will be illustrated using specific parameters of existing products. SiC power devices with a voltage rating of 1 200 V and a current rating of approximately 25 A at the 100 ℃ time of selection, as well as commonly used Si power devices with the same voltage and current rating, have been selected for reference illustration and their parameters are listed in Tab. 11. 2. 2.

Tab. 11. 2. 2　Main parameters of 1 200 V SiC and Si power devices

Category	Si MOSFETs	Si IGBT	SiC MOSFETs	SiC JFET (off)	SiC JFET (on)
Model	ATP12040L2FLL	APT25GR120B	CMF2010D	SJEP120R063	IJW120R070T1
Packaging	TO264	TO247	TO247	TO247	TO247
Manufacturer	Microsemi	Microsemi	CREE	semiSouth	Infineon
Rated voltage U_{ds}/V	1 200	1 200	1 200	1 200	1 200
Rated current I_{ds}/A(25/100 ℃)	30 (25 ℃)	75	42	30	36
On-resistance $R_{ds(on)}$/mΩ	400	100	80	63	70
Operating temperature T_j/℃	−55—150	−55—150	−55—135	−55—150	−55—175

(1) On-Resistance Characteristics

The on-state resistance characteristics of SiC and Si power devices under different driving conditions is shown in Fig. 11. 2. 3.

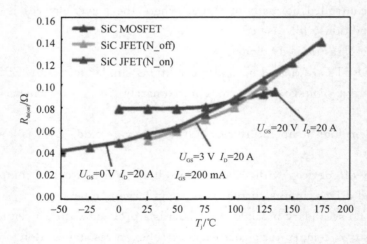

Fig. 11. 2. 3　On-state resistance characteristics of SiC and Si power devices
under different driving conditions

The temperature characteristics of the on-state resistance of SiC and Si power devices are shown in Fig. 11. 2. 4.

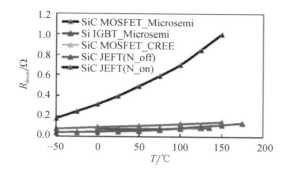

Fig. 11. 2. 4   Temperature characteristics of the on-states resistance of SiC and Si power devices

Fig. 11. 2. 4 shows that SiC MOSFETs $R_{ds(on)}$ have a lower temperature coefficient and are less temperature sensitive $R_{ds(on)}$ than SiC JFETs.

The on-state resistance of SiC power devices is much lower than that of Si power devices at high voltage levels of 1 200 V. This shows that SiC power devices have lower on-state losses for high voltage applications and can increase energy conversion efficiency with higher breakdown voltages.

(2) Threshold Voltage Characteristics

The threshold voltage characteristics of SiC and Si power devices are shown in Fig. 11. 2. 5. The same type of power devices, the threshold voltage of SiC power devices is less than that of Si power devices; the threshold voltage of normally broken SiC JFETs is lower than that of SiC MOSFETs, which is only about 1. 0 V. The design of the driver circuit needs to prevent other signals from interfering with the false turn-on of power devices, especially in high-frequency switching applications, where the faster turn-off speed is likely to lead to a larger ringing phenomenon, thus causing the false turn-on of power devices. In high-frequency switching applications, faster turn-off speeds can easily lead to larger ringing phenomena, resulting in false turn-on of the power devices.

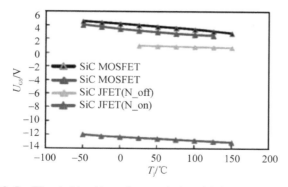

Fig. 11. 2. 5   Threshold voltage characteristics of SiC and Si power devices

In addition, both SiC JFETs and SiC MOSFETs have a negative temperature coefficient of threshold voltage, and measures need to be taken to ensure reliable shutdown of the power devices when used in high temperature applications.

　　(3) Transfer Characteristics

　　The transfer characteristics of SiC and Si power devices are shown in Fig. 11. 2. 6.

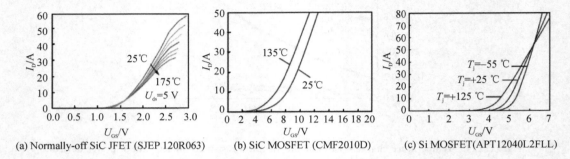

(a) Normally-off SiC JFET (SJEP 120R063)　　(b) SiC MOSFET (CMF2010D)　　(c) Si MOSFET(APT12040L2FLL)

Fig. 11. 2. 6　Transfer characteristics of SiC and Si power devices

　　As can be seen in Fig. 11. 2. 6, the saturation drain current of the SiC JFET has a negative temperature coefficient over almost the entire operating range (1. 5 $V \leqslant U_{GS} \leqslant 3. 0$ V) and therefore has a good inherent resistance to transient current surges.

　　SiC MOSFETs have a negative temperature coefficient for saturated drain current at $U_{GS} \geqslant$ 15 Varound and a positive temperature coefficient for saturated drain current at $U_{GS} <$ 15 V around.

　　Si MOSFETs have a negative temperature coefficient for saturated drain current at $U_{GS} \geqslant 6 \sim 7$ V; and a positive temperature coefficient for saturated drain current at $U_{GS} < 6 \sim 7$ V.

　　SiC MOSFETs and Si MOSFETs have a positive temperature coefficient of saturation drain current when the drive voltage is low and operating in the linear operating region, when a short circuit shock will more easily lead to thermal failure of the device, which is not conducive to parallel and safe use of power devices. In contrast, SiC JFETs are more suitable for applications where current shock situations need to be handled frequently.

　　(4) Anti-avalanche breakdown capability of the power devices

　　Tab. 11. 2. 3 lists the avalanche breakdown resistance of SiC and Si power devices.

Tab. 11. 2. 3　Avalanche breakdown resistance of SiC and Si power devices

Category	Si MOSFETs			SiC MOSFETs	
Model	ATP12040L2FLL	IPW90R120C3	R6076ENZ1	CMF2010D	SCT30N120
Manufacturer	Microsemi	Infineon	Rohm	CREE	STM
Maximum single pulse avalanche energy EAS/J	3. 2	1. 94	1. 95	2. 2	1. 0
Maximum repetitive pulse avalanche energy EAR/mJ	50	2. 9	2. 96	1. 5	—
Test conditions	—	$I_D = 8. 8$ A $U_{DD} = 50$ V	$I_D = 13. 4$ A $U_{DD} = 50$ V	$I_D = 20$ A $U_{DD} = 50$ V	$I_D = 20$ A $U_{DD} = 150$ V

As listed in Tab. 11. 2. 3, SiC can withstand a maximum single pulse avalanche energy $EAS(SiC) \approx EAS(Si)$; maximum repetitive pulse avalanche energy $EAR(SiC) \approx EAR(Si)$, thus in terms of overall performance, SiC power devices have more outstanding resistance to avalanche breakdown compared to Si power devices.

(5) Switching Time and Switching Loss Characteristics

The specific parameters for switching times and switching losses for SiC and Si power devices are listed in Tab. 11. 2. 4 and Tab. 11. 2. 5 respectively.

Tab. 11. 2. 4   Switching losses for SiC and Si power devices in the 1 200 V voltage class

Category	Model	Manufacturers	Conduction loss/μJ	Turn-off loss/μJ	Test conditions
Si MOSFETs	APT12040L12FLL	Micro Semi	265	147	$U_{DD}=600$ V,$I_D=30$ A
Si IGBT	APT25GR120B	Micro Semi	742	427	$U_{DD}=600$ V,$I_D=25$ A
SiC MOSFETs	CMF2010D	CREE	110	150	$U_{DD}=800$ V, $I_D=10$ A
SiC JFET(off)	SJEP120R063	Semi South	131	222	$U_{DD}=600$ V,$I_D=24$ A
SiC JFET(on)	IJW120R070T1	Infieon	340	210	$U_{DD}=800$ V, $I_D=10$ A

Tab. 11. 2. 5   Switching times for SiC and Si power devices in the 1 200 V voltage class

Category	Model	Manufacturers	On time/ns	Off time/ns	Test conditions
Si MOSFETs	APT12040L12FLL	Micro Semi	35	91	$U_{DD}=600$ V,$I_D=30$ A
Si IGBT	APT25GR120B	Micro Semi	26	142	$U_{DD}=600$ V,$I_D=25$ A
SiC MOSFETs	CMF2010D	CREE	64	78	$U_{DD}=600$ V,$I_D=20$ A
SiC JFET(off)	SJEP120R063	Micro Semi	27	65	$U_{DD}=600$ V,$I_D=24$ A

From the relevant data in Tab. 11. 2. 4 and Tab. 11. 2. 5, it can be seen that SiC power devices at 1 200 V voltage level have a greater reduction in total switching losses and switching times than Si power devices.

SiC power devices are more suitable for high-voltage and high-frequency applications, helping to increase switching frequency, reduce system losses and reduce the size and cost of passive components in converters. A comparison of the main characteristics of SiC and Si power devices shows that SiC power devices have outstanding performance advantages over Si power devices at high voltage levels; normally-off SiC JFETs have outstanding performance and higher reliability (e. g. negative temperature coefficient of saturation drain current, higher temperature reliability, etc.) compared to SiC MOSFETs and commonly used Si MOSFETs and Si IGBTs.

11.3   The Application of SiC Power Devices in Aviation

11.3.1   Summary

In aviation and aerospace, the power conversion and distribution units proposed in the US third generation MEA programme mainly use SiC power devices. The special characteristics of multiple energy ions in the aerospace application environment have irradiation resistance requirements for equipment and devices. SiC, GaN and other WBG semiconductor material power devices have very good application prospects because of their excellent irradiation resistance characteristics.

Starting from the needs of MEA, the French aero-engine manufacturing company Safran has developed a 2 kW inverter combining Si IGBT and SiC SBD, which reduces switching losses by 45% and increases efficiency. The University of Nottingham in the UK has applied the SiC SBD to an aerospace matrix inverter, also achieving a reduction in switching losses, an increase in switching frequency and an increase in system efficiency.

Due to the development trend of high-power and high-voltage PSSs in aviation and aerospace, mechanical DC circuit breakers are prone to arcing in the switching process, especially under high-voltage conditions, with low reliability and life, and high noise interference, usually to avoid pulling arcs and take vacuum arc suppression devices. For workplaces with limited space, such as aviation, aerospace, armoured vehicles and ships, the operating environment is complex and harsh, requiring DC circuit breakers to be small in size, light in weight, highly reliable, long-life and low-noise. The use of arc-free and contact-free solid-state protection switches with high life and reliability is better suited to the needs of the aerospace sector. Solid-state protective switches based on SiC power devices are also valuable for research as they have lower on-resistance and better high-temperature resistance than Si under high-voltage conditions and are more suitable for high-temperature, high-power and high-reliability applications.

SiC and GaN power transistors have gradually begun to be applied and have brought significant performance improvements, and new broadband power semiconductor devices are bound to be widely used in the future.

Based on the advantages of SiC power devices, it is necessary to analyze the application of SiC in aviation secondary power sources and the impact of overall performance.

11.3.2   Problems in SiC Power Device Applications

1. Issues in the Development of New WBG SiC Power Devices

(1) Limitations of the Production Process

The application of SiC power devices in high temperature, high reliability power electronics devices is severely affected by the limitations of the production process such as epitaxy, doping, trenching and packaging, as well as the long-term operational stability of the gate oxide layer in SiC MOSFETs. The performance of existing commercial SiC power devices should be weighed against the performance requirements of the system to select the right device.

(2) Higher Production Costs

SiC power devices must use SiC material as a substrate, making the cost of SiC power devices much higher than Si power devices, which is also one of the main reasons limiting the popularity of SiC power devices in the field of consumer power electronics, and how to reduce production costs by improving wafer size and qualification rate is also a problem that needs to be further studied and solved in the future.

2. Focus on Research for SiC Power Devices in Power Electronics

(1) Analysis and Modelling study of SiC Power Device Characteristics

SiC power devices have different characteristics from Si power devices due to the large differences in the properties of SiC and Si materials. The modelling of power devices can help to provide in-depth analysis of device characteristics, simplify the design process and optimise the application of SiC devices.

(2) Design of Drive Circuits for SiC Power Devices

SiC power devices must be considered for drive compatibility when replacing Si power devices in mature power electronics devices. Existing commercially available SiC MOSFETs are well compatible with Si MOSFET/IGBT applications, as long as they meet the performance requirements of the device. Some normally-broken SiC JFETs require sufficient drive current and a narrow range of drive voltage control, necessitating further design of the drive circuit in power electronics devices where precise control of the drive voltage value is required. IXYS has developed a more mature driver chip for SiC JFETs.

(3) Research of High Frequency Magnetic Elements

SiC power devices are suitable for high frequency applications and can significantly improve system performance. As power devices make it possible to increase the switching frequency of the system, the system also places higher demands on the performance of magnetic components, such as high-frequency transformers and power inductor elements: high frequency and high reliability. However, both the selection of core materials, the core production and manufacturing process, as well as the structural design and performance optimization of transformers and inductors, are yet to be studied in depth.

(4) Topology Optimization and Innovation

The use of SiC power devices is aimed at improving system performance and meeting the requirements of harsh environments, and how to achieve the outstanding performance benefits of SiC power devices through topology optimization of power electronics devices is a major concern.

(5) Assembly Study of SiC Power Devices

SiC power devices can be used in high-temperature, high-voltage, high-power applications, meeting the increasing demand for power capacity in power systems, such as aerospace vehicles. However, the increase in system power capacity also brings with it the problem of thermal design of high-power devices. How to optimize the thermal structure of the system through thermal design, improve the heat dissipation capacity and reduce the weight and volume of the heat dissipation is a concern and a necessary condition to ensure safe and reliable operation of the devices and equipment.

(6) The High Temperature Performance Study of the Other Components

Although the maximum junction temperature of SiC rate devices cannot reach the theoretical high temperature level due to the packaging process, research has shown that SiC power devices can be used in high temperature applications on bare die. To ensure that the entire system can operate at high temperatures, the other components of the device also need to be able to withstand the corresponding harsh environment, so improving the temperature resistance of the other components of the system is also a concern for researchers.

The application of SiC power devices has many advantages, such as greatly improved device performance, reduced losses, reduced size and weight, simplified heat dissipation conditions and resistance to irradiation. But at the same time, it also brings many new problems and challenges, such as the high-reliability distribution switch with arc extinguishing function for high-voltage DC networks, the manufacturing and sealing process of high-temperature resistance, the optimization of the drive circuit and topology, and the improvement of the performance requirements of other components of the system, etc. Whether SiC can be widely used in the aerospace field will also be constrained by its high production cost, but once its production cost is effectively reduced, SiC power devices will rapidly become popular in all fields of power electronics. Once the production cost is effectively reduced, SiC power devices will be rapidly popularized in all related fields of power electronics.

Based on the advantages of SiC power devices, we analyse the application of SiC in aerospace secondary power supplies and the impact brought about by the performance of the complete machine. Compared to Si devices, power electronic converters made from SiC power devices will achieve an increase in overall machine performance. Here, a typical converter is used as an example to present the performance improvement obtained.

11.3.3 Application of SiC Devices in Aerospace DC/AC Converters

DC/AC converters convert DC to AC, e. g. static converters, motor drives, etc.

1. Applications in Aeronautical Static Converters

The Aeronautical Static Inverter (ASI) is a secondary power source that converts DC voltage to AC for supplying emergency AC loads on board when the primary power supply fails. The ASI is a static converter of low voltage 28 V or high voltage 270 V DC into single or three phase AC with an output voltage of 115 V/200 V, 36 V or 26 V and a frequency of 400 Hz using power semiconductor devices. A typical block diagram of a sinusoidal pulse width modulated two-stage static converter is shown in Fig. 11. 3. 1. The first stage is a DC/DC stage that converts 270 V DC to 180V DC and achieves electrical isolation, and the second stage is a DC/AC stage that inverts 180 V DC to 115 V/400 Hz.

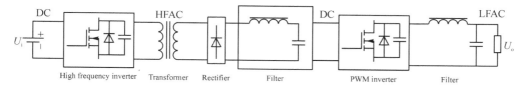

Fig. 11. 3. 1 Typical block diagram of a sinusoidal pulse width modulated two-stage static converter

The front-stage converter is used as the DC input link of a bridge inverter with an input voltage $U_i = 270$ V and an output voltage of 180 V or more, and is electrically isolated. DC/DC converter topologies such as two-device forward, push-pull, half-bridge and full-bridge are often used. Half-bridge, full-bridge circuit and two-device forward, push-pull circuit compared to the main switch device to bear the voltage stress reduced by half, after considering the safety margin, should be selected to withstand the voltage of about 500 V power devices. For this voltage withstand level, the on-state resistance of an ordinary Si device is about 1 Ω, while the on-state resistance of a SiC device is only a few tens of milliohms, and its on-state loss is theoretically only one percent of that of a Si device, so that the system efficiency is improved by using SiC devices. Therefore, SiC Schottky diodes with high withstand voltage are required.

The efficiency and power density of the rear stage inverter are key indicators of the power converter. An inverter with an input DC voltage of 360 V, an output AC voltage of 230 V/50 Hz and a rated output power of 1 000 V · A was chosen for analysis. A comparison of the efficiency of different devices at different frequencies is shown in Fig. 11. 3. 2. Three different device combinations of all-Si devices, hybrid devices (combination of Si power devices and SiC diodes) and all-SiC devices were used respectively.

As can be seen from Fig. 11. 3. 2, with all-Si devices, the switching losses increase the most and the efficiency decreases the most as the switching frequency increases. For the lower switching frequency of 23. 4 kHz, the difference in efficiency between the all-Si device inverter and the all-SiC device inverter is not significant, at 1. 8%, but when the switching frequency is increased to 100 kHz, the difference in efficiency is greater, with the all-SiC device inverter being 8% more efficient than the all-Si device inverter.

2. Applications in Electric Motor Drives

The DC/AC type motor drive converts 270 V high voltage DC into variable frequency AC power for the motor and drives the motor. Multi-compatible motors also require a DC/AC power converter. 1 fault-tolerant motor is equivalent to 2 motors operating in backup or parallel to each other, suitable for use in aircraft. For example, in many aircraft there are two electric fuel pumps in the fuel tank. If one of them fails, there is still fuel going to the

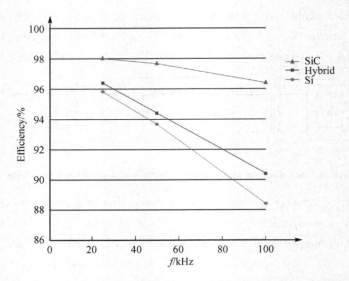

Fig. 11.3.2　Comparison of the efficiency of different devices at different frequencies

engine. With a fault-tolerant motor, only one electric fuel pump is needed in one tank. Although fault tolerant motors are not yet used on aircraft, this is a definite trend.

Each phase of the fault tolerant motor is powered by a mono-phase DC/AC converter, and each converter is powered by a different DC supply, so that whether one phase of the motor fails, or one converter fails or one power supply fails, the motor can still run continuously, even without reducing output power. A six-compatible motor can be powered by two three-phase DC/AC converters. The advantage of this solution is that fewer switching device are used.

The DC/AC converter for fault tolerant motors operates in SPWM sinusoidal pulse width modulation mode to obtain sinusoidal phase currents so that the magnetic field of the motor is circular or elliptical to reduce the iron and copper losses of the motor. The phase windings of the fault tolerant motor are isolated from each other and are not connected to the housing, so the DC/AC converter does not need to be isolated either.

The use of SiC devices in motor drive systems is being investigated due to their superior characteristics. The categories are SiC/Si hybrid devices, all-SiC-JFETs, all-SiC-MOSFETs and all-SiC-BJT motor drives.

11.3.4　Application of SiC Devices in Transformer Rectifiers

In a PSS where the main power supply is AC, a DC secondary power supply is essential in order to supply power to on-board electronic equipment such as radio communications, radar, cockpit equipment, control and protection devices, relays, signalling devices and DC power equipment such as motors. In addition to the requirement for small size and weight, the converter must have high efficiency. According to the type of transformer in the rectifier, they can be divided into Transformer Rectifier Unit (TRU) and Auto Transformer

Rectifier Unit (ATRU).

1. Transformer Rectifier Unit

The AC/DC converter converts the inverted AC power to DC power, which is commonly used in conventional aircraft for conversion by a voltage converter TRU, such as the 12 – pulse voltage converter TRU structure. As shown in Fig. 11.3.3, it is the schematic diagram of the 12 – pulse voltage converter. The 12-pulse voltage converter consists of a three-phase transformer with star input and star and triangle output, which converts the three-phase AC power through a two-way three-phase bridge rectifier circuit. The output DC voltage of the two-way bridge rectifier is output to the load via an inter phase reactor L_P.

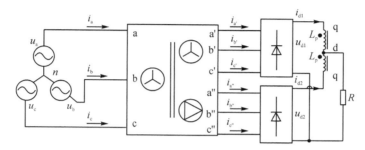

Fig. 11.3.3   12-Pulse transformer rectifier unit

To obtain lower harmonics and a higher power factor, this can be achieved by increasing the number of transformers and rectifiers in the TRU. For example, an 18 – pulse TRU contains three transformers and three rectifier bridges. However, weight and size are a big issue in aircraft and TRUs higher than 18 pulses are not used.

The B787 PSS is a practical multi-pulse transformer rectifier used to rectify 230 V AC bus voltage to 28 V low voltage DC output to supply load devices such as DC fuel pumps, ignition, cockpit displays, etc.

2. Auto Transformer Rectifier Unit

In an MEA inverter AC power system, the main power supply is 230 V inverter AC power with a frequency range of 360—800 Hz. For the convenience of the load, it is usually converted to high voltage DC power by an AC/DC converter, e. g. ±270 V DC or 540 V DC or other voltage values applicable to the after power equipment, mainly using an ARTU to achieve AC/DC power conversion for the MEA's electric hydraulic pumps, nitrogen generation systems, environmental control compressors, environmental control cooling fans and engine start-ups.

The ATRU is a high power factor AC/DC passive rectifier conversion using a combination of phase shift transformers and diode rectifiers, while the PWM rectifier is an active high power factor rectifier conversion. As shown in Fig. 11.3.4, it is a symmetrical 12-pulse auto transformer rectifier.

However, the ATRU can only transmit power in one direction and cannot absorb the

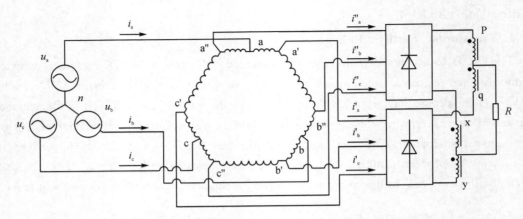

Fig. 11. 3. 4   Symmetrical 12 - pulse auto transformer rectifier

braking regenerative energy of the speed control motor, but can only consume energy with resistance, resulting in lower efficiency. In the long term, unless higher performance magnetic materials are available, the power density of ATRUs will be difficult to increase significantly.

Fig. 11. 3. 3 and Fig. 11. 3. 4 show TRU and ATRU. Because the main component of the rectifier circuit is a diode, its voltage stress is the maximum value of the input line voltage of the rectifier bridge. For multi-pulse transformer rectifiers, generally choose diodes with a voltage tolerance of several hundred volts and a current quota of tens to hundreds of amps, so ordinary Si diodes can meet the requirements. While for multi-pulse auto transformer rectifiers, fast recovery diodes with a voltage tolerance of 600 V and a current rating of tens to hundreds of amperes are generally used, which have reverse recovery problems. The switching speed of SiC MOSFETs and SiC JFETs is much higher than that of Si CoolMOS and Si IGBT, which can significantly reduce switching losses. At the same time, the high junction temperature capability of SiC devices allows the switching frequency to be increased further, thus reducing the size and overall weight of passive components such as magnetic components and capacitors.

11.3.5 SiC Devices in DC/DC Converter

DC/DC converters convert one type of DC power to another voltage, and DC/DC converters are widely used in aircraft. The GCU, the generator controller in the aircraft power system, requires a DC power supply. The internal power supply of the Constant Speed Constant Frequency (CSCF) power and Variable Speed Constant Frequency (VSCF) power controllers. GCU uses ordinary buck and buck/boost converter. The input voltage of these converters is the DC output voltage after rectification and filtering of the permanent magnet auxiliary exciter, and the output voltage of the converters is used by modules with different functions within the GCU.

Avionics equipment requires DC power supply, with two types of DC input voltages:

low voltage DC with a rated voltage of 28 V DC, and 270 V DC power supply. 270 V DC is mainly used for transmitters in communication, radar, electronic countermeasures equipment, and various cathode ray tube displays. The low-voltage outputs include 24 V DC, 12 V DC, 5 V DC, 6.3 V DC, and 3.3 V DC, which are used in various low-voltage circuits.

Typical DC/DC converters include buck, boost and push-pull types (push-pull, half-bridge and full-bridge converters). In applications where the input and output voltages are low, mostly modular power supplies are used. Where the input is high voltage and the output is high or low voltage, mostly isolated direct-to-direct converters are used.

To investigate the performance of SiC, a non-isolated buck converter was chosen for the study, as shown in Fig. 11.3.5.

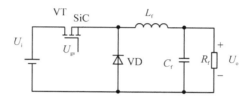

Fig. 11.3.5　Simplified non-isolated buck circuit diagram

SiC-MOSFETs were used for the main power device and SiC fast recovery diodes and SiC Schottky diodes were used for the renewal diode. The heat sink temperatures of the SiC-MOSFETs and renewal diode were measured, as shown in Fig. 11.3.6, which compares the heat sink temperatures when Si-based diodes and SiC Schottky diodes are used, and the measurement results reflect that using SiC Schottky diode as a current-continuity diode results in a lower temperature rise. This is mainly due to the SiC Schottky diode's ability to significantly reduce the losses in the system, resulting in lower temperatures for the MOSFETs and the current-continuation diode.

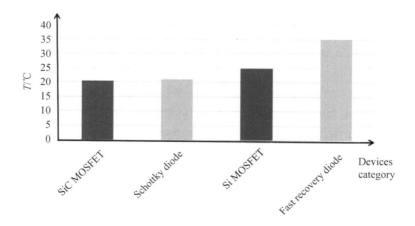

Fig. 11.3.6　Heat sink temperature when using Si-based diodes compared to SiC Schottky diodes

As shown in Fig. 11. 3. 7, for the resonant forward circuit schematic, when the input voltage $U_i = 270$ V DC, the output voltage $U_o = 28$ V DC, the selection of the primary side of the main power device need to take into account the relationship between the maximum blocking voltage and the resonant peak, assuming that the resonant voltage peak is 800 V. Taking into account the safety margin, it is necessary to use the maximum blocking voltage of 1 200 V power device.

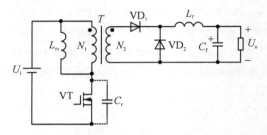

Fig. 11. 3. 7　Resonant forward circuit schematic

Under this quota, Si-MOSFETs have a higher on-resistance and Si-IGBTs have a current dragging problem, which limits the switching frequency. SiC-MOSFETs have a smaller on-resistance and can work at high frequencies, which is more suitable for use. The theoretical maximum voltage of the secondary rectifier diode is about 200 V, while the theoretical maximum voltage of the continuity diode is 85 V, so the two need to use different diodes, rectifier choose 400 V/40 A, and the continuity device choose 200 V/20 A. At this time, if the ordinary Si diode is used, its reverse recovery time is generally tens of nanoseconds, which will affect the efficiency of the whole machine, so that the heat increases. If it is replaced by SiC Schottky diode, it can make the system temperature reduced, and make efficiency increased.

11.3.6　SiC Devices for AC/AC Conversion and Motor Drive Applications

The flight control actuators, fuel pump motors and fans in environmental control systems in aircraft all require motor drives with increasing power requirements and higher reliability requirements in addition to size, weight and efficiency. For example, according to reliability studies of electro-mechanical actuators in MEAs, multi-phase drives offer greater advantages in terms of fault tolerance than three-phase drive topology, making AC/AC converters an integral part of motor drives to improve reliability capabilities.

1. Fault-tolerant Drive Forms for Electric Motors

A motor multiphase drive is the implementation of a multiphase excitation of the motor winding, where the number of phases of the motor is usually 5, 6 and 7, etc. A model of a multiphase drive with a three – phase input, n – phase output converter is shown in Fig. 11. 3. 8, where the AC/AC converter is connected in an n – phase star pattern.

It has been shown that multi-phase topologies do not meet the fault-tolerant operation required for MEA applications. In order to achieve true fault tolerance, two methods are available for MEA multi-phase motor drives, i. e. a multi-AC/AC mono-phase motor fault-tolerant drive architecture shown in Fig. 11. 3. 9 and a multi-AC/AC

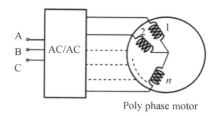

Fig. 11. 3. 8　Multi-phase AC/AC drive motor

three-phase motor fault-tolerant drive architecture shown in Fig. 11. 3. 10.

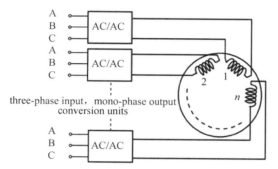

Fig. 11. 3. 9　Fault-tolerant drive architecture for multiple AC/AC mono-phase motors

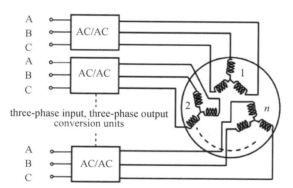

Fig. 11. 3. 10　Fault-tolerant drive architecture for multiple AC/AC three-phase motors

2. Electric Motor Drive Circuit Forms

There are three solutions to the motor drive fault tolerance problem, namely the Voltage DC link Back-to-Back Converters (V – BBCs), the Indirect Matrix Converter (IMC) and the Direct Matrix Converter (DMC).

(1) Voltage DC Link Back-to-Back Converters

The V-BBC is divided into two parts, rectifier and inverter, as shown in Fig. 11. 3. 11. The first stage is a three-phase controlled rectifier output with a capacitor C as an energy storage element, while the second stage is the inverter stage, which converts DC power into AC power. Both stages of the circuit are independent into closed-loop control.

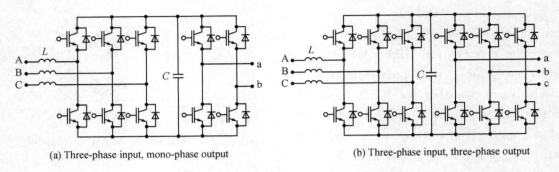

(a) Three-phase input, mono-phase output   (b) Three-phase input, three-phase output

Fig. 11. 3. 11   Fault-tolerant multiphase drive for V-BBC units

Both three-phase input/mono-phase output and three-phase input/three-phase output V-BBC construction circuits use controllable switching devices and diodes. For example, the three-phase input/three-phase output V-BBC has 12 controllable switches, 12 diodes and a three-phase input inductor.

(2) Indirect Matrix Converter

The indirect matrix converter topology is shown in Fig. 11. 3. 12, where Fig. 11. 3. 12 (a) is a three – phase input, mono-phase output converter and Fig. 11. 3. 12(b) is a three – phase input, three – phase output converter composition, requiring 18 controlled switching devices and 18 diodes.

The switching devices in the circuit are also subjected to the same peak voltage as the V-BBC, so the power switching devices used in the IMC have the same voltage rating as the power switches in the V-BBC.

In contrast, the average withstand voltage or the effective value of the withstand voltage of the power switching device is approximately half that of the corresponding device in the V-BBC circuit. As a result, the voltage stress of the IMC, i. e. the ratio between the applied voltage and the rated voltage, is reduced by approximately half with respect to the V-BBC.

From a control point of view, the maximum output voltage is 86. 6% of the effective value of the input voltage. Compared to the V-BBC, IMC control is more complex due to the increased number of switching devices. A control scheme of space vector modulation is often used, where the rectifier stage current and the output stage voltage are space vector modulated respectively.

(3) Direct Matrix Converter

The Direct Matrix Converter(DMC), shown in Fig. 11. 3. 13, is a three-phase input, three-phase output AC/AC scheme with no intermediate DC link. The DMC uses the same number of power switching devices as the IMC circuit.

The output voltage is 86. 6% of the RMS value of the input voltage and the control of the AC/AC converter is very similar to that of an IMC, requiring multiple switching devices and more complex digital logic, usually implemented using Field Programmable Gate Array

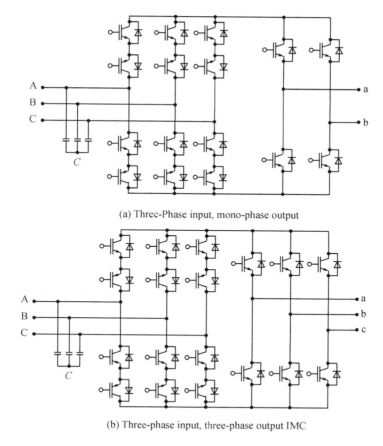

(a) Threc-Phase input, mono-phase output

(b) Three-phase input, three-phase output IMC

Fig. 11. 3. 12   Indirect matrix converter toplogy

(FPGA) devices.

3. Power Switching Device Performance Improvement Has Become an Urgent Need

With the increase in the number of MEA motors, the reliability requirements for fault tolerance, energy storage etc. have increased and the number of power switching devices and diodes in the converter has increased, so the improvement in the performance of the power switching devices has had a significant effect on the overall performance.

The high temperature resistance and almost zero reverse recovery current of the SiC Schottky diodes greatly improve the performance of motor drives, reduce dissipated power, size and weight and increase product reliability.

In addition, when the SiC MOSFET process has matured, it will be possible to further reduce the size and weight of motor drives due to their high temperature performance, if they can replace the switching power devices currently in use.

Tab. 11. 3. 1 and Tab. 11. 3. 2 are used for a motor drive, comparing the efficiency and losses at different temperatures at two frequencies of 10 kHz and 20 kHz for three different

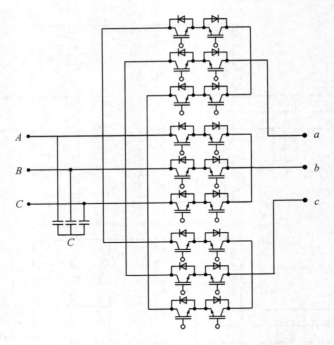

Fig. 11. 3. 13　AC/AC direct converters

device combinations of all-Si devices, hybrid devices and all-SiC devices respectively.

Tab. 11. 3. 1　Efficiency of inverters for motor drives

$T/℃$　　　f/kHz　　$n/\%$	10			20		
	Si	Mixed	SiC	Si	Mixed	SiC
70	95. 10	96. 00	97. 10	92. 20	93. 82	95. 70
105	94. 50	95. 80	97. 10	90. 70	93. 52	95. 60

Tab. 11. 3. 2　Inverter losses for motor drives

kJ

$T/℃$　　　Name	Si	Mixed	SiC	Si	Mixed	SiC
70	771. 68	627. 52	458. 57	1 280. 7	997. 23	681. 69
105	879. 42	663. 72	468. 22	552. 3	1 049. 0	69. 80

From Tab. 11. 3. 1, it can be seen that at a switching frequency of 10 kHz the efficiency of the Si inverter drops by 0. 6% at 105 ℃ compared to 70 ℃, however due to the high temperature resistance of the SiC inverter the efficiency drops by only 0. 1% and for the hybrid inverter the efficiency drops by 0. 2%. When the switching frequency is increased from 10 kHz to 20 kHz, the Si inverter efficiency drops by 3. 1% at 70 ℃ and 3. 8% at 105 ℃, while the SiC inverter efficiency drops by only 1. 4%. These experimental results have initially reflected the superior high-temperature and high-frequency characteristics of SiC devices. As the manufacturing technology of SiC devices further improves and matures, the advantages of SiC devices and their use in the whole machine will surely be expanded even more.

Most of the losses in energy conversion are consumed in power conversion and power drive. In the face of increasing energy constraints, reducing system energy consumption and developing new energy sources has become an urgent need for sustainable development.

11.3.7   Application of SiC in SSPC

With the introduction of advanced MEA and AEA, technological advances in the field of power electronics and the increasing demand for electrical equipment in current aircraft, the requirements of aircraft power distribution systems for energy-efficient aviation have been raised to a new level of development.

Traditional aircraft structures with mechanical, hydraulic and pneumatic systems have inherent drawbacks such as heavy weight, large size, inflexible infrastructure, long response times and difficulty in locating faults. The trend towards MEA/AEA requires the replacement of traditional mechanical, hydraulic and pneumatic loads with electrical loads to improve basic performance, reliability and maintainability. Replacing conventional aircraft systems with electrical systems can significantly improve aircraft reliability, maintainability and performance in terms of weight, size and system complexity.

The installation of a large number of new devices in existing aircraft, the need to connect them to the new electrical system and the increasing need for conversion circuits such as converters and inverters as the voltages of the new devices must coexist with conventional voltages, all mean that new aircraft power distribution systems are becoming increasingly large and complex. In the current automatic aircraft power distribution system, power electronics-based devices play a fundamental role, with solid-state power controllers playing a pivotal role in the automatic aircraft power distribution system as part of line protection and electrical load control. As shown in Fig. 11. 3. 14, it is the structural framework of the automatic power distribution system of an aircraft.

A typical SSPC structure diagram is shown in Fig. 11. 3. 15. The SSPC consists of a power circuit, a control circuit, and a communication circuit. The power circuit is mainly composed of a power switch, a current detection section, and an external power device. Generally, MOSFETs or IGBTs are used as power switches to turn on/off control of the power circuit. Current detection is often achieved by using a sampling resistor or current sensor connected in series in the power circuit. If the current is greater than the set maximum current, the drive circuit of the power device is quickly switched off to achieve short circuit protection; otherwise, the circuit is protected by a delay according to the inverse time curve. The digital intelligent SSPC can be used with software to achieve time delay protection and to provide the SSPC with a thermal memory function. The SSPC uses isolation circuits and communication interfaces to transfer control commands and status signals to the control computer.

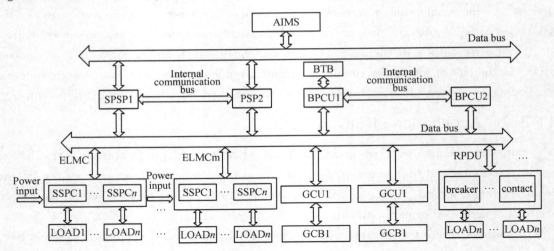

Fig. 11. 3. 14　Structure framework of the automatic power distribution system of an aircraft

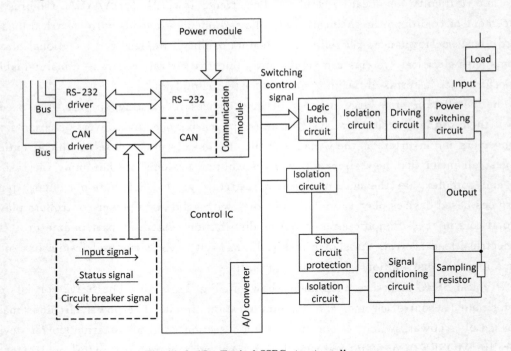

Fig. 11. 3. 15　Typical SSPC structure diagram

Solid-state power controllers exhibit superior performance to relays and circuit breakers. The introduction of SiC switching devices will further enhance the performance of solid-state power controllers in three ways.

(1) Faster Response

The switching devices of SiC have better high frequency performance and the SSPC response time of applied SiC devices is less than 1 μs.

(2) Higher Power Ratings and Reliability

SiC devices can operate at higher temperatures, at higher voltages, and SiC SSPCs can have higher power ratings and higher reliability.

(3) Higher Power Density

The high operating temperature reduces the bulk mass of the heat sink, even without active cooling, and increases the power density of the SSPC.

Due to the low permissible junction temperature of the silicon base, the operation is dependent on the cooling of the loop control system. For example, the high-power power electronics of the B787, the ATRU and the DC/AC converter of the speed-controlled motor have been unable to adapt to air-cooling and have switched to water-cooling cooling methods, increasing the complexity of the cooling system. As the number of power electronics for MEA increases, the power increases, so if its efficiency and power density can be further improved, it will certainly promote the development of MEA electrical systems, which in turn will enable the aircraft to reduce weight, reduce fuel consumption and improve aircraft performance.

The on-resistance and parasitic capacitance of power electronics devices determine the conduction losses and switching losses of power electronics devices respectively. Their losses are both a major part of the overall power electronics losses, largely affecting the efficiency of the power electronics, and are also one of the most significant sources of heat in power electronics. The characteristics and parameters of the different models of power electronics differ. The level of switching frequency determines the losses generated in the converter and therefore the maximum switching frequency that can be allowed varies. The level of switching frequency determines the volume and weight of the passive components in the converter and largely influences the power density of the power electronics. In addition, the high rates of voltage and current change caused by increased switching frequencies can also lead to serious EMI problems.

The voltage and current tolerance of power electronics devices and heat dissipation performance determines the failure rate of devices, and device failure is an important factor in the reliability of power electronics devices; the ability of power electronics devices to work at high temperatures can reduce the heat dissipation requirements of power electronics devices, and help reduce the volume and weight of cooling devices, improve the power density of power electronics devices, and adapt to harsh high temperature operating environments.

At present, it is difficult to significantly reduce the on-resistance and junction capacitance of silicon devices, which makes it difficult to significantly reduce their conduction and switching losses, limiting the efficiency of Si-based power converters (referred to as Si-based converters). Even with soft switching technology, the switching frequency can be increased to a certain extent, but this increases the complexity of the circuit and has a negative impact on reliability. High and very high speed motors based on air and magnetic float bearings are one of the key components of the electrical system of MEA. The maximum junction temperature that a typical Si device can withstand is 150 ℃. Even with the latest processes and sophisticated liquid-cooled heat dissipation techniques, Si devices struggle to break 200 ℃ operating temperature, which is far from adequate for the requirements of MEA.

In general, silicon devices after nearly 60 years of development, its performance has been close to the theoretical limits of its material limits, it is difficult to significantly improve performance through device structure innovation and process improvement. Power electronic devices based on new WBG semiconductor materials have become an important development trend.

In the last 50 years, power electronics has been widely used in aerospace vehicles and has contributed to the development of aviation science and technology. However, due to the rapid development of aerospace technology itself, new requirements have been put forward for power electronics devices and devices, i. e. the development of high temperature power electronics devices and high temperature integrated circuits must be accelerated in order to in turn improve system reliability and reduce the size and weight of power electronics devices; low pass-state voltage drop devices, including diodes and transistors, must be developed. They are not only to be used in solid-state power controllers, they are of great value in realizing the uninterrupted power supply of DC PSSs, low voltage output switching power supplies and modular power supplies as well as motor drive speed regulation and other links that are indispensable.

11.4 Summary

In recent years, aircraft manufacturers have been replacing pneumatic and hydraulic systems with electrical energy in order to reduce CO_2 emissions and contribute to compliance with green environmental requirements, in addition to the requirements of the use of electrical equipment in terms of maintenance, power control, traceability and efficiency. Aircraft electrical equipment becomes an integral part of the total mass of the aircraft, and in order to reduce the mass of electrical equipment, system level configuration and the adoption of new technologies (use of new devices such as gallium nitride, silicon carbide, etc.) as well as research into topology, or the integration of existing components, are key issues to be addressed by the aircraft industry.

Power electronics is one of the key technologies for the realization of MEA. In addition to the topology being researched at length, electronic components, magnetic elements and capacitors are constraining the development of power electronics and the development of high temperature power electronics is imminent.

Power electronics are used in a wide range of applications in MEA, such as various AC/DC, DC/DC and DC/AC converters, solid state power controllers, transformer rectifiers, motor drives and MEEs.

The use of SiC technology greatly reduces losses compared to Si technology and is resistant to high temperatures, which is very conducive to increasing power density, reducing volume and weight and improving reliability, etc. , and is conducive to promoting the development of aerospace electrical technology.

11.5 Exercises

1. What challenges are the frequency conversion AC power supply system facing?

2. Why do the power electronic converter based on Si devices have performance bottlenecks?

3. What are the SiC characteristics and advantages over Si?

4. What are the remaining problems of SiC in the application of power device?

5. What are the SiC applications of power devices in the aviation field?

Appendix

Tab. A. 1 Acronyms and Abbreviation

SN	English name	Chinese name	ACR. and ABB.
1	Air Bypass Valve	空气旁通阀门	ABV
2	Actuator Control Electric Unit	作动器控制电子装置	ACEU
3	Alternator Current Exciter	交流励磁机	ACE
4	Air Cycle Machine	空气循环机	ACM
5	Audio Control Panel	音频控制面板	ACP
6	AC Power Supply	交流电源	ACPS
7	All Electric Aircraft	全电飞机	AEA
8	Alkaline Fuel Cell	碱性燃料电池	AFC
9	Audio Gateway Unit	音频网关组件	AGU
10	Air Heating Valve	空气加热阀	AHV
11	Active Magnetic Bearing	有源磁轴承	AMB
12	Aircraft Maintenance Manual	飞机维修手册	AMM
13	ARINC664 network Cabinet Switch	ARINC 664 网络机柜交换机	ACS
14	ARINC664 network Switch	ARIN 664 网络交换机	ANS
15	Auxiliary Power	辅助供电	AP
16	All Power Aircraft	全动力飞机	APA
17	Auxiliary Power Breaker	辅助电源断路器	APB
18	Auxiliary Power Control Relay	辅助电源控制继电器	APCR
19	Auxiliary Power Control Unit	辅助动力控制单元	APCU
20	Auxiliary Power Unit	辅助动力装置	APU
21	ARINC664 network Remote Switch	ARINC 664 网络远程交换机	ARS
22	APU Starter/Generator	APU 启动发电机	ASG
23	Air Turbine	空气涡轮发动机	AT
24	Auto Transformer Rectifier Unit	自耦合变压整流单元	ATRU
25	Air Valve System	空气阀门系统	AVS
26	Buck/Boost Converter Unit	降压/升压变换器	BBCU
27	Brake Command	刹车指令	BC
28	Battery Charger Regulation Unit	蓄电池充电调节装置	BCRU
29	Brushless Direct Current Motor	无刷直流电动机	BLDCM
30	Brake Operation Test	制动操作测试	BOT
31	Brake Position Command	刹车位置指令	BPC
32	Bus Power Control Unit	汇流条功率控制单元	BPCU
33	Brake Power System Controller	刹车电源系统控制器	BPSC
34	Brake System Control Unit	刹车系统控制单元	BSCU
35	Burner Staging Valve	燃烧分段阀门	BSV

(**Continued**)

SN	English name	Chinese name	ACR. and ABB.
36	Bus Tie Breaker	汇流条连接断路器	BTB
37	Cabin Air Compressor	座舱空气压缩机	CAC
38	Controller Area Network	控制器区域网络	CAN
39	Cabin Attendant Panel	客舱乘务员面板	CAP
40	Circuit Breaker	电路断路器	CB
41	Common Computing Resource	公共计算资源	CCR
42	Common Computer System	公共计算机系统	CCS
43	Common Core System	共同核心系统	CCS
44	Common Data Network	公共数据网络	CDN
45	Cabin Electrical Air Compressor	客舱电气空气压缩机	CEAC
46	Constant Frequency Power Supply System	恒频电源系统	CFPSS
47	Constant Frequency Generator	恒频发电机	CFG
48	Central Maintenance Computing Function	中央维护计算功能	CMCF
49	Common Motor Start Controller	通用电动机启动控制器	CMSC
50	Current Return Network	电流回馈网络	CRN
51	Constant Speed Device	恒速驱动装置	CSD
52	Cabin Services System Controller	座舱服务系统控制器	CSSC
53	Current Transformer	电流互感器	CT
54	Cabin Zone Unit	客舱区域组件	CZU
55	Direct Current	直流	DC
56	Display and Crew Alert Function	显示和机组告警功能	DCAF
57	Electric Air Compressor	电动空气压缩机	EAC
58	Electrical Bus	电气总线	EB
59	Electric Brake Actuator Controller	电刹车作动控制器	EBAC
60	Electric Brake Power Supply Unit	电动制动器电源装置	EBPSU
61	Emergency Condition	紧急情况	EC
62	Environment Control System	环境控制系统	ECS
63	Economic Cooling Valve	经济冷却阀	ECV
64	Electronic and Electrical Controller	电子电气控制器	EEC
65	Electronic Flight Bag	电子飞行包	EFB
66	Electro-Hydraulic Servo Valve	电液伺服阀	EHSV
67	Engine Indication and Crew Alerting System	发动机指示和机组警告系统	EICAS
68	Electrical Load Management Center	电气负载管理中心	ELMC
69	Electric Mechanical Actuator	电机械作动器	EMA
70	Electro-magnetic Compatibility	电磁兼容	EMC
71	Electro-magnetic Interference	电磁干扰	EMI

(Continued)

SN	English name	Chinese name	ACR, and ABB.
72	External Power	外电源	EP
73	External Power Relay	外电源继电器	EPR
74	Emergency Power Unit	应急动力装置	EPU
75	Electrical Redundant Power Distribution	电气冗余功率分配	ERPD
76	Full Authority Digital Engine Control System	全权数字发动机控制系统	FADECS
77	Flight Control Computer	飞行控制计算机	FCC
78	Flight Control Electronics	飞行控制电子	FCE
79	Flight Control Electric Actuator	飞行控制电作动器	FCEA
80	Flying Control System Power Supply	飞控系统供电电源	FCSPS
81	Flight Control Management Cabinet	飞行控制管理机柜	FCMC
82	Fan Shaft Driven Emergency Power Supply System	风扇轴驱动应急电源系统	FSDEPSS
83	Flight Management Computer	飞行管理计算机	FMC
84	Flight Management Function	飞行管理功能	FMF
85	Fuel Management Unit	燃油管理装置	FMU
86	Fiber Optical Bus	光纤总线	FOB
87	Fiber Optic Converter	光纤转换器	FOC
88	Fan Shaft Drive Generator	风扇轴驱动发电机	FSDG
89	Generator Breaker	发电机断路器	GB
90	Generator Control	发电机控制器	GC
91	Generator Control Unit	发电机控制单元	GCU
92	Graphics Generator	图形发生器	GG
93	General Processor Module	通用处理模块	GPM
94	Hot Battery Bus	热电池汇流条	HBB
95	Hybrid Excitation Synchronous Motor	混合励磁同步电机	HESM
96	High Flow Shut Off Valve	高流量关断阀	HFSOV
97	High-Energy Laser Weapon	高能激光武器	HELW
98	High Intensity Discharge	高亮度	HID
99	Hydraulic Power	液压能	HP
100	High Pressure Starter/Generator	高压启动发电机	HPSG
101	High Pressure Turbine Cooling Control	高压涡轮冷却控制	HPTCC
102	High Voltage Alternative Current	高压交流电	HVAC
103	High Voltage Direct Current	高压直流	HVDC
104	Hydraulic Interface Function	液压接口功能	HIF
105	Integrated Actuator Package	组合式作动器组件	IAP
106	Integrated Cooling System	集成冷却系统	ICS
107	Integrated Drive Generator	组合传动发电机	IDG

SN	English name	Chinese name	ACR. and ABB.
108	In Flight Entertainment	飞行娱乐系统	IFE
109	Inlet Guide Vane	进气口导叶片	IGV
110	Indirect Matrix Converter	间接矩阵变换器	IMC
111	Integrated Power Unit	整体功率单元	IPU
112	Load Controller	负载控制器	LC
113	Liquid Crystal Display	液晶显示器	LCD
114	Light Emitting Diode	发光二极管	LED
115	Low-pressure Electric Fuel Pump	低压电动燃油泵	LEFP
116	Landing Gear Actuation	起落架作动器	LGA
117	Local Hydraulic Central	局部液压中心	LHC
118	Low Pressure Limit Valve	低压限制阀门	LPLV
119	Line Replaceable Unit	现场可更换单元	LRU
120	Linear Variable Differential Transformer	线性差动变压器	LVDT
121	Low Pressure Turbine Cooling Control	低压涡轮冷却控制	LPTCC
122	Landing Gear Indication and Steering	起落架指示和转向装置	LGIS
123	Main Busbar	主汇流条	MBB
124	Matrix Converters	矩阵变换器	MC
125	Molten Carbonate Fuel Cell	熔融碳酸盐燃料电池	MCFC
126	More-Electric Aircraft	多电飞机	MEA
127	More-Electric Engine	多电发动机	MEE
128	Medium-pressure Electric Fuel Pump	中压电动燃油泵	MEFP
129	Mechanical Electronic Power Controller	机械电子功率控制器	MEPC
130	More-Electric Propulsion System	多电推进系统	MEPS
131	More Electric Full Authorized Digital Electronic Controller	多电全权数字电子控制器	MEFADEC
132	Main Landing Gear	主起落架	MLG
133	Main Minimum Equipment List	主最低设备清单	MMEL
134	Multi-Power Aircraft	多动力飞机	MPA
135	Mean Time Between Failure	平均无故障间隔时间	MTBF
136	Nitrogen Generation System	氮气发生系统	NGS
137	Nose Landing Gear	前起落架	NLG
138	Negative Pressure Safety Valve	负压安全活门	NPSV
139	Non Return Valve	止回阀,单向阀门	NRV
140	Nose Wheel Steering	前轮转向装置	NWS
141	Nose Wheel Steering System	前轮转向系统	NWSS

(Continued)

SN	English name	Chinese name	ACR. and ABB.
142	On Board Inertia Gas Generator System	机载惰性气体发生系统	OBIGGS
143	On Board Maintenance System	机载维护系统	OBMS
144	Over Current	过流	OC
145	Outflow Valve	出风阀	OFV
146	Outflow Valve Controller	外流阀控制器	OFVC
147	Phosphoric Acid Fuel Cell	磷酸燃料电池	PAFC
148	Power Conditioning Module	电源调节模块	PCM
149	Pack Controller Unit	组件控制器	PCU
150	Passenger Control Unit	乘客控制组件	PCU
151	Power Electric Cooling System	功率电子冷却系统	PECS
152	Power Electronic Module	功率电子模块	PEM
153	Polymer Electrolyte Fuel Cell	聚合物电解质燃料电池	PEFC
154	Proton Exchange Membrane Fuel Cell	质子交换膜燃料电池	PEMFC
155	Power Electric Cooling System	功率电子冷却系统	PECS
156	Primary Electrical Power Distribute Central	初级电气功率分配中心	PEPDC
157	Primary Flight Control	主飞行控制	PFC
158	Power Factor Correct	功率因数校正	PFC
159	Permanent Magnet Generator	永磁发电机	PMG
160	Permanent Magnet Synchronous Motor	永磁同步电动机	PMSM
161	Power Optimization Aircraft	功率优化飞机	POA
162	Passenger Oxygen System	乘客氧气系统	POS
163	Pneumatic Power Source	气压能	PPS
164	Positive Pressure Safety Valve	正压安全活门	PPSV
165	Passenger Service Module	乘客服务模块	PSM
166	Propulsion Thrust Power	推进能量	PTP
167	Phase Sequence	相序	PS
168	Passenger Sign	乘客标志	PS
169	Power Supply System	电源系统	PSS
170	Pulse Width Modulator	脉冲宽度调制	PWM
171	Ram Air Turbine	冲压式空气涡轮	RAT
172	Remote Data Concentrate	远程数据集中器	RDC
173	Remote Electronic Unit	遥控电子单元	REU
174	Remote Power Distribute Unit	远程功率分配单元	RPDU
175	Rejected Take Off	拒绝起飞	RTO
176	Remote Sensor Unit	远程传感器	RSU
177	Rotary Variable Differential Transformer	旋转差动器变压器	RVDT

(**Continued**)

SN	English name	Chinese name	ACR. and ABB.
178	Surge Control Valve	浪涌控制阀	SCV
179	Secondary Distribute Power System	二次配电系统	SDPS
180	Secondary Electrical Power Distribute Central	次级电气配电中心	SEPDC
181	Starter/Generator	启动发电机	S/G
182	State Of Charge	电池的荷电状态	SOC
183	Solenoid Valve	螺线管阀	SV
184	Solenoid Operated Pilot Valve	电磁阀操作试验阀	SOPV
185	Shut Off Valve	关断阀	SOV
186	Switch Reluctance Generator	开关磁阻发电机	SRG
187	Solid State Power Controller	固态功率控制器	SSPC
188	Total Harmonic Distortion	总谐波失真度	THD
189	Totally Integrated More-Electric System	完全综合多电系统	TIMES
190	Throttle Level Angle	油门杆位置角度	TLA
191	Tiller Module	操纵手柄模块	TM
192	Thrust Resolver Angle	推力解析器角度	TRA
193	Transfer Rectifier Unit	变压整流单元	TRU
194	Variable Bleed Valve	可变排气阀	VBV
195	Value Control Unit	阀控组件	VCU
196	Variable Exhaust Valve	可变排气阀	VEV
197	Variable Frequency Power Supply System	变频电源系统	VFPSS
198	Voltage Regulator	电压调节器	VR
199	Variable Speed Constant Frequency	变速恒频	VSCF
200	Variable Stator Vane	变距静叶	VSV
201	Wear Pin	磨损针	WP
202	Wide Band Gap	宽带能隙	WBG
203	Wound Rotor Synchronous Machine	线绕转子同步机	WRSM

References

［1］WILD T W. Transport Category Aircraft System［M］. Englewood：Jeppesen 2008.

［2］Moir I, SEABRIDGE A. Aircraft systems mechanical, electrical, and avionics subsystems integration［M］. West Sussex：John Wiley & Sons, Ltd, 2008.

［3］CHAYINTHU W, AREERAK K, AREERAK K. The Dynamic Model of Electrical Aircraft System Feeding a Buck — Boost Converter［C］//5th International Electrical Engineering Congress, Pattaya, Thailand, 8—10 March 2017.

［4］EISMIN K T. Aircraft electricity and electronics［M］. 6th ed. New York：McGraw-Hill Education, 2014.

［5］CORCAU J I , DINCA L. More Electricity on Modern Civil Aircrafts-review［C］// 2020 24th International Conference on Circuits, Systems, Communications and Computers (CSCC), 2020.

［6］SARLIOGLU B, MORRIS C T. More Electric Aircraft：Review, Challenges, and Opportunities for Commercial Transport Aircraft［J］. IEEE TRANSACTIONS ON TRANSPORTATION ELECTRIFICATION, 2015,1(1):54-64.

［7］MADONNA V, GIANGRANDE P, Galea M. Electrical Power Generation in Aircraft：Review, Challenges, and Opportunities ［ J ］. IEEE TRANSACTIONS ON TRANSPORTATION ELECTRIFICATION, 2018,4(3):646-659.

［8］ZHAO X, GUERRERO J M, WU X H. Review of Aircraft Electric Power Systems and Architectures［C］// ENERGYCON 2014, May 13-16, Dubrovnik, Croatia, 2014.

［9］YANG Z, QU J, Ma Y, et al. Modeling and simulation of power distribution system in more electric aircraft［J］. Journal of Electrical and Computer Engineering, 2015(1):1-7.

［10］DENG D. Li-ion batteries：basics, progress, and challenges［J］. Energy Science & Engineering, 2015, 3(5):385-418.

［11］CHAKRABORTY I, TRAWICK D, JACKSON D, et al. Electric Control Surface Actuator Design Optimization and Allocation for the More Electric Aircraft［C］// AIAA Aviation, August 12-14, 2013, Los Angeles, CA, 2013 Aviation Technology, Integration, and Operations Conference,2013.

［12］Boeing Company. Aircraft Maintenance Manual Chapter 21 Air Conditioning ［Z］, 2010.

［13］Boeing Company. 787-8 Training Student Notebook Block Five - Part B［Z］, 2010.

［14］Boeing Company. 787-8, Training Student Notebook Block Five - Part C［Z］, 2010.

［15］Boeing Company. 787 Training Student Lab Notebook Wheels & Brakes and Nose

Wheel Steering System - Hydraulics Book 4 of 4[Z], 2010.

[16] SARLIOGLU B, MORRIS C T. More electric aircraft: review, challenges, and opportunities for commercial transport aircraft [J] IEEE Trans. Transportation Electrification, 2015,1(1):54-64,.

[17] ZHAO X, GUERRERO J M, WU X. Review of aircraft electric power systems and architectures[C]// in Proc. IEEE International Energy Conf. , 2014:949-953.

[18] SINNETT M. 787 No-Bleed Systems: saving fuel and enhancing operational efficiencies [J]. Boeing Aero, 2007(4): 6-11.